POWER QUALITY ISSUES

CURRENT HARMONICS

Accessing the E-book edition

Using the VitalSource® ebook

Access to the VitalBook™ ebook accompanying this book is via VitalSource® Bookshelf – an ebook reader which allows you to make and share notes and highlights on your ebooks and search across all of the ebooks that you hold on your VitalSource Bookshelf. You can access the ebook online or offline on your smartphone, tablet or PC/Mac and your notes and highlights will automatically stay in sync no matter where you make them.

1. **Create a VitalSource Bookshelf account at** *https://online.vitalsource.com/user/new* or log into your existing account if you already have one.

2. **Redeem the code provided in the panel below to get online access to the ebook.** Log in to Bookshelf and click the **Account** menu at the top right of the screen. Select **Redeem** and enter the redemption code shown on the scratch-off panel below in the **Code To Redeem** box. Press **Redeem**. Once the code has been redeemed your ebook will download and appear in your library.

DOWNLOAD AND READ OFFLINE

To use your ebook offline, download BookShelf to your PC, Mac, iOS device, Android device or Kindle Fire, and log in to your Bookshelf account to access your ebook:

On your PC/Mac

Go to *http://bookshelf.vitalsource.com/* and follow the instructions to download the free **VitalSource Bookshelf** app to your PC or Mac and log into your Bookshelf account.

On your iPhone/iPod Touch/iPad

Download the free **VitalSource Bookshelf** App available via the iTunes App Store and log into your Bookshelf account. You can find more information at *https://support.vitalsource.com/hc/en-us/categories/200134217-Bookshelf-for-iOS*

On your Android™ smartphone or tablet

Download the free **VitalSource Bookshelf** App available via Google Play and log into your Bookshelf account. You can find more information at *https://support.vitalsource.com/hc/en-us/categories/200139976-Bookshelf-for-Android-and-Kindle-Fire*

On your Kindle Fire

Download the free **VitalSource Bookshelf** App available from Amazon and log into your Bookshelf account. You can find more information at *https://support.vitalsource.com/hc/en-us/categories/200139976-Bookshelf-for-Android-and-Kindle-Fire*

N.B. The code in the scratch-off panel can only be used once. When you have created a Bookshelf account and redeemed the code you will be able to access the ebook online or offline on your smartphone, tablet or PC/Mac.

SUPPORT

If you have any questions about downloading Bookshelf, creating your account, or accessing and using your ebook edition, please visit *http://support.vitalsource.com/*

POWER QUALITY ISSUES
CURRENT HARMONICS

SURESH MIKKILI • ANUP KUMAR PANDA

CRC Press
Taylor & Francis Group
Boca Raton London New York

CRC Press is an imprint of the
Taylor & Francis Group, an **informa** business

CRC Press
Taylor & Francis Group
6000 Broken Sound Parkway NW, Suite 300
Boca Raton, FL 33487-2742

First issued in paperback 2020

© 2016 by Taylor & Francis Group, LLC
CRC Press is an imprint of Taylor & Francis Group, an Informa business

No claim to original U.S. Government works

ISBN 13: 978-0-367-57558-8 (pbk)
ISBN 13: 978-1-4987-2962-8 (hbk)

Dedicated to my better half Ms. Alekhya.

"What you are today is God's gift to you, and what

you become tomorrow is your gift to God."

—Dr. Suresh Mikkili

Contents

Preface

Electronic equipment such as computers, battery chargers, electronic ballasts, variable-frequency drives, and switched-mode power supplies generate perilous harmonics and cause enormous economic loss every year. Because of that, both power suppliers and power consumers are concerned about power quality problems and compensation techniques. Harmonics surfaced as a buzzword in the 1980s and threatened the normal operation of power systems and user equipment. Harmonics issues are of great concern to engineers and building designers because they can do more than distort voltage waveforms; they can overheat a building's wiring, cause nuisance tripping, overheat transformer units, and cause random end-user equipment failure. Thus, power quality (PQ) has continued to become a more serious issue. As a result, active power filters (APFs) have gained much more attention due to excellent harmonic and reactive power compensation in two-wire (single phase), three-wire (three-phase without neutral), and four-wire (three-phase with neutral) AC power networks with nonlinear loads.

Active power filters have been under research and development for more than three decades and have found successful industrial applications with varying configurations, control strategies, and solid-state devices. However, this is still a technology under development, and many new contributions and new control topologies have been reported in the last few years. It is aimed at providing a broad perspective on the status of APF technology to researchers and application engineers dealing with power quality issues.

In Chapter 1, the importance of active power filters and solid-state devices is explained in detail, and APF configurations and selection considerations of them are also presented.

In Chapter 2, proportional–integral (PI) controller–based shunt active filter (SHAF) control strategies (p-q and I_d-I_q) are discussed in detail. SHAF control strategies for extracting three-phase reference currents are compared, with their performance evaluated under different source voltage conditions using a PI controller. The performance of the control strategies has been evaluated in terms of harmonic mitigation and DC link voltage regulation. The detailed simulation results are presented to support the feasibility of proposed control strategies. To validate the proposed approach, the system is also implemented on real-time digital simulator hardware, and adequate results are reported for its verification.

In Chapter 3, type 1 fuzzy logic controller (FLC)–based SHAF control strategies with different fuzzy membership functions (MFs) (trapezoidal, triangular, and Gaussian) are developed for extracting three-phase reference currents, and are compared by evaluating their performance under different source voltage conditions. The performance of the control strategies has been evaluated in terms of harmonic mitigation and DC link voltage regulation. Detailed simulation and real-time results are presented to validate the proposed research.

Even though type 1 FLC–based SHAF control strategies with different fuzzy MFs are able to mitigate the harmonics, notches are presented in the source current. So to mitigate the harmonics perfectly, one has to choose a perfect controller. Therefore, in Chapter 4, type 2 FLC–based SHAF control strategies with different fuzzy MFs (trapezoidal, triangular, and Gaussian) are introduced. With this approach, the compensation capabilities of SHAF are extremely good. The detailed simulation results using MATLAB®/ Simulink® software are presented to support the feasibility of the proposed control strategies.

In Chapter 5, a specific class of digital simulator known as a real-time simulator is introduced by answering the questions "What is real-time simulation?" "Why is it needed?" and "How does it work?" The latest trend in real-time simulation consists of exporting simulation models to a *field-programmable gate array* (FPGA). Today every researcher wants to develop his or her model in real time. The steps involved for implementation of a model from MATLAB to real time are provided in detail. The proposed type 2 FLC– based SHAF control strategies with different fuzzy MFs are verified with a real-time digital simulator (OPAL-RT) to validate the proposed research.

Last, Chapter 6 summarizes the book and looks at future work. A comparative study of PI controllers and the proposed type 1 FLC– and type 2 FLC– based SHAF control strategies with different fuzzy MFs using MATLAB and a real-time digital simulator is also presented.

MATLAB is a trademark of The MathWorks, Inc. and is used with permission. The MathWorks does not warrant the accuracy of the text or exercises in this book. This book's use or discussion of MATLAB software or related products does not constitute endorsement or sponsorship by The MathWorks of a particular pedagogical approach or particular use of the MATLAB software. For product informaton, please contact:

The MathWorks, Inc.
3 Apple Hill Drive
Natick, MA 01760-2098 USA
Tel: 508 647 7000
Fax: 508-647-7001
E-mail: info@mathworks.com
Web: www.mathworks.com

Acknowledgments

It has been a pleasure to work on this book. I hope the reader will find it not only interesting and useful, but also comfortable to read.

I am thankful to the director of the National Institute of Technology, Goa, India (Prof. G. R. C. Reddy) and the director of the National Institute of Technology, Rourkela, India (Prof. S. K. Sarangi), for their inspiring guidance, unmatched support, and blessings.

I express my sincere gratitude to my academic and research advisor Prof. Anup Kumar Panda for his excellent guidance, research attitude, encouragement, and kind cooperation throughout the course of my research work. His timely help and painstaking efforts made it possible to present the work contained in this book. I consider myself fortunate to have worked under his guidance.

I am obliged to the faculty and staff of the Electrical and Electronics Engineering Department of the National Institute of Technology, Rourkela, and the National Institute of Technology, Goa.

I express my heartfelt thanks to international journal reviewers for giving their valuable comments on published papers in different international journals, which helps to carry the research work in the right direction. I also thank international conference organizers for their in-depth reviewing of the published papers.

I express my deep sense of gratitude and reverence to my beloved father, Mikkili Kantha Rao; mother, Suvarna Pushpa; brother, Seshu Babu; and sister, Sowjanya, and my dear wife Alekhya who supported and encouraged me the entire time, no matter what difficulties I encountered. I express my greatest admiration to all my family members and relatives for the positive encouragement they showered on me throughout this research work. Without my family's sacrifice and support, this research work would not have been possible. I am especially indebted to my well-wishers, Sampathy and Jayanthi, for their continuous support and encouragement.

It is a great pleasure for me to acknowledge and express my appreciation to all my well-wishers for their understanding, relentless support, and encouragement during my research work. Last but not least, I express my sincere thanks to all those who directly or indirectly helped me at various stages of this work.

Heartfelt appreciation is also due to the editorial and production staff of CRC Press of India for their excellent contributions toward the production values of the book.

Above all, I thank the Almighty God for the wisdom and perseverance that he has bestowed upon me during this research work and, indeed, throughout my life.

What you are today is God's gift to you, and what you become tomorrow
is your gift to God.

Dr. Suresh Mikkili
Department of Electrical and Electronics Engineering
National Institute of Technology–Goa

Authors

Dr. Suresh Mikkili is an assistant professor and head of the Electrical and Electronics Engineering Department, National Institute of Technology, Goa, India. He earned his PhD (2010–2013) and master's (MTech) (2006–2008) in electrical engineering from the National Institute of Technology, Goa, and his BTech (2002–2006) in electrical and electronics engineering from SITE (Sasi Institute of Technology and Engineering), T. P. Gudem, affiliated with JNT (Jawaharlal Nehru Technological) University, Andhra Pradesh, India. He worked as an assistant professor in electrical engineering in distinguished engineering colleges from June 2008 to July 2010. He is a professional member of the Institute of Electrical and Electronics Engineers.

Dr. Mikkili's main area of research includes power quality improvement issues, active filters, power electronic applications to power systems, and applications of soft computing techniques. He has delivered several lectures in his research area. He has reported results of his research (30+ articles) in reputed international journals (such as *IET Power Electronics*, *IJEPES* (*International Journal of Electrical Power and Energy Systems*), *ETEP* (*European Transactions on Electrical Power*), *FEE* (*Frontiers of Electrical Engineering*), the *Journal of Power Electronics*, the *Journal of Electrical Engineering*, and *IJEEPS*) and international conferences (such as IEEE-IECON, Canada; IEEE-PEDS, Singapore; IEEE-PES-ISGT, the United States; IEEE-INDICON-BITS PILANI, Hyd; IEEE-SCES-MNIT, Allahabad; IEEE-ICPS-IIT, Madras; and IEEE-PEDES-IISc, Bangalore).

Anup Kumar Panda earned his BTech in electrical engineering from Sambalpur University, India; MTech in power electronics and drives from the Indian Institute of Technology, Kharagpur, India; and PhD from Utkal University in 1987, 1993, and 2001, respectively. In 1990, he joined IGIT (Indria Gandhi Institute of Technology), Sarang, as a lecturer serving there for 11 years, and then in January 2001, he joined the National Institute of Technology, Rourkela, as an assistant professor and currently is continuing as a professor in the Department of Electrical Engineering. He has published more than 85 articles in journals and conferences. He has completed two MHRD (Ministry of Human Resource Development) projects and one NaMPET (National Mission of Power Electronics Technology) project, as well as guided six PhD scholars. Presently, he is guiding ten scholars in the area of power electronics and drives. He is a fellow of the Institute of Engineering and Technology, UK; Institute of Engineers, India; and Institute of Electronics and Telecommunication Engineering. He is also a senior member of the Institute of Electrical and Electronics Engineers, United States.

Dr. Panda's research interests include the design of high-frequency power conversion circuits and applications of soft computing techniques, and improvements in multilevel converter topology, power factors, and power quality in power systems and electric drives.

Abbreviations

ADC	Analog-to-digital converter
AMD	Advanced microdevice
APF	Active power filter
APLCs	Active power line conditioners
APQCs	Active power quality conditioners
ARTEMIS	Advanced real-time electro-mechanical transient simulator
ASDs	Adjustable-speed drives
BJT	Bipolar junction transistor
BOA	Bisector of area
COA	Centroid of area
COTS	Commercial off the shelf
CS	Console subsystem
CSI	Current source inverter
CT	Current transformer
DAC	Digital-to-analog converter
DSP	Digital signal processor
EEPROM	Electrically erasable programmable read-only memory
EMI	Electromagnetic interference
EMT	Electromagnetic transient
EMTP-RV	Electromagnetic Transients Program—Restructured Version
FIS	Fuzzy inference system
FLC	Fuzzy logic controller
FOU	Foot point of uncertainty
FPGA	Field-programmable gate array
GTO	Gate-turn-off thyristor
GUI	Graphical user interface
HDL	Hardware description language
HIL	Hardware-in-the-loop
HVDC	High-voltage direct current
I_d-I_q theory	Instantaneous active and reactive current theory
IEEE	Institute of Electrical and Electronics Engineers
IGBT	Insulated gate bipolar transistor
INTEL	Integrated electronics
IRPC	Instantaneous reactive power compensator
JTAG	Joint Test Action Group
LMF	Lower membership function
LOM	Largest (absolute) value of maximum
MATLAB	Matrix Laboratory

MF	Membership function
MOM	Mean value of maximum
MOSFET	Metal-oxide semiconductor field effect transistor
MS	Master subsystem
NB	Negative big
NM	Negative medium
NS	Negative small
PB	Positive big
PCI Express	Peripheral Component Interconnect Express
PE	Power electronics
PI	Proportional–integral
PLC	Programmable logic controller
PM	Positive medium
PQ	Power quality
p-q theory	Instantaneous active and reactive power theory
PS	Positive small
PSB	Power system blockset
PSCAD	Power systems computer aided design
PSS/E	Power system simulator for engineering
PT	Potential transformer
PWM	Pulse width modulation
RTOS	Real-time operating system
RTSI	Real-time system integration
RTW	Real-time workshop
SHAF	Shunt active filter
SIT	Static induction thyristor
SOM	Smallest (absolute) value of maximum
SRF	Synchronous reference frame
SS	Slave subsystem
TCP/IP	Transmission Control Protocol/Internet Protocol
THD	Total harmonic distortion
TNA	Transient network analyzer
Type 1 FLC	Type 1 fuzzy logic controller (T1FLC)
Type 2 FLC	Type 2 fuzzy logic controller (T2FLC)
UMF	Upper membership function
UPQC	Unified power quality conditioner
UPS	Uninterruptible power supply
VSI	Voltage source inverter
ZE	Zero

Notations

A	Fuzzy set
$\tilde{A}$	Type 2 fuzzy set
c	Center
E	Error
ΔE	Change in error
i_c	Compensation/filter current
i_{ca}, i_{cb} and i_{cc}	Compensation/filter currents of phases A, B, and C
i_{cn}	Compensation/filter current of neutral
i_{ca^*}, i_{cb^*} and i_{cc^*}	Reference current of phases A, B, and C
$i_{c\alpha ref}$ (or) $i_{c\alpha^*}$	∞-axis reference current
$i_{c\beta ref}$ (or) $i_{c\beta^*}$	β-axis reference current
i_{c0ref}	Zero-axis reference current
i_l	Load current
i_{la}, i_{lb} and i_{lc}	Load current of phases A, B, and C
i_{ln}	Load current of neutral
i_{ld}	d-axis load current
i_{lq}	q-axis load current
$i_{l\alpha}$	∞-axis load current
$i_{l\beta}$	β-axis load current
i_s	Source current
i_{sa}, i_{sb} and i_{sc}	Source current of phases A, B, and C
i_{sn}	Source current of neutral
i_0	Zero-axis load current
i_{0p} and i_{0q}	Instantaneous zero-sequence active and reactive currents
$i_{\alpha p}$ and $i_{\alpha q}$	Instantaneous active and reactive currents on α-axis
$i_{\beta p}$ and $i_{\beta q}$	Instantaneous active and reactive currents on β-axis
m	Fuzzification factor
P_0	Zero-sequence instantaneous real power
$P_{\alpha\beta}$	Instantaneous real power due to +*ve* and −*ve* sequence components
i_{Ld1h} and i_{Lq1h}	Fundamental frequency components of i_{ld} and i_{lq}
σ	Width
$\mu_A(x)$	Membership function
$\mu_{\tilde{A}}(x,u)$	Type 2 membership function
V_a, V_b and V_c	Voltages across phases A, B, and C to neutral
V_{dc}	Actual DC link voltage
V_{dc-ref}	Reference DC link voltage
V_{dc1} and V_{dc2}	DC link voltages of split capacitors 1 and 2
v_α	∞-axis voltage
v_β	β-axis voltage
X	Universe of discourse

1

Introduction

1.1 Research Background

Recent advancements in power electronics has encouraged large scale use of the non-linear loads such as adjustable speed drives (ASD), traction drives, etc. Power electronics based power converters have contributed for the deterioration of the power quality and this has resulted in the increase of power losses and great economic loss. The electronic equipment like computers, battery chargers, variable frequency drives, etc., generate harmonics and cause great economic loss every year. The concept of harmonics was noticed in 1980's and since then it has been a major problem in the power system. Harmonics can do much more than distort voltage waveforms, they can overheat the building wiring, overheat transformer units, etc. This made it important to develop equipment that can mitigate the harmonics present in the power system and we call them as harmonic Filters.

Power Quality (PQ), is defined as "Any power problem manifested in voltage, current or frequency deviation which leads to damage, malfunctioning and disoperation of the consumer equipment." Poor power quality causes many damages to the system, and has a negative economic impact on the utilities and customers. Highly automatic electric equipment, in particular, causes enormous economic loss every year. The problems of harmonics can be reduced or mitigated by the use of power filters.

In the early days, passive filters were used for mitigation of power quality problems in general and harmonics in particular. But their performance is limited due to the problem of resonance and lack of dynamic compensation. Active power filters have been developed to overcome these limitations. Active power filters have been proven very effective in the reduction of the system harmonics. One of the most severe and common power quality problems is current harmonics. In particular, voltage harmonics and power distribution equipment problems result from current harmonics.

The voltage generated at the generating station is not purely sinusoidal. Due to the nonuniformity of the magnetic field and the winding distribution in a working AC machine, voltage waveform distortions are created, and thus the voltage obtained is not purely sinusoidal. The distortion at the point of generation is very small (about 1% to 2%), but still it exists.

When a pure sinusoidal voltage is applied to a certain type of load, the current drawn by the load is proportional to the voltage and the current waveform is similar to voltage waveform. These loads are referred to as linear loads (loads where the voltage and current follow one another without any distortion to their pure sine waves). Examples of linear loads are resistive heaters, incandescent lamps and constant speed induction motors. In contrast, some loads cause the current to vary disproportionately with the voltage during each half cycle. These loads are defined as nonlinear loads. The current harmonics and the voltage harmonics are generated because of these nonlinear loads. It is to be noted that the non-sinusoidal current results in many power system problems such as low-power factor, low-energy efficiency, electromagnetic interference (EMI), power system voltage fluctuations and so on.

Various standards and guidelines have been established to specify the limits on the magnitudes of harmonics currents and voltages. Institute of Electrical and Electronics Engineers (IEEE) specify the limits on the voltages at various harmonics frequencies. In 1983 IEEE working group made a reference about harmonic source and effects on the electric power system. Christopher recognized the harmonic related problems and started work on a standard that would give guidelines to users and engineers and according to those guidelines IEEE 519 was formulated in 1981. In 1996,IEEE working group proposed definitions for power terms that are practical and effective when voltage and current are distorted or unbalanced. It also suggested definitions for measurable values that may be used to indicate the level of distortion and unbalance. The IEEE standard 1459 is intended to evaluate the performance of modern equipment or to design and build the new generations of instrumentation for energy and power qualification.

Conventionally, passive filters are available for the elimination of harmonics. However, these L-C filters introduce tuning, aging, resonance problems and these filters are large in size and are suited for fixed harmonic compensation. This is the reason why power engineers and researchers are working on the dynamic,complete and comprehensive solution to PQ problems. The solution is provided by active power filters (APFs), also known as active power line conditioners. They are capable of suppressing various PQ issues such voltage and current compensation, reactive power compensation, voltage flicker problems, system unbalance problems, etc.

In this book, the performance of the shunt active filter (SHAF) current control strategies has been evaluated in terms of harmonic mitigation and DC link voltage regulation. This research presents different control strategies and controllers with enhanced performance of shunt active filters for power quality improvement by mitigating the harmonics and maintaining a constant DC link voltage. Three-phase reference current waveforms generated by proposed schemes are tracked by the three-phase voltage source converter in a hysteresis band control scheme.

For extracting the three-phase reference currents for shunt active power filters, we have developed instantaneous active and reactive power p-q and instantaneous active and reactive current I_d-I_q control strategies. For regulating and maintaining a constant DC link capacitor voltage, the active power flowing into the active filter needs to be controlled. In order to maintain a constant DC link voltage and generate the compensating reference currents, we have developed proportional–integral (PI), type 1, and type 2 fuzzy logic controllers with different fuzzy membership functions (MFs) (trapezoidal, triangular, and Gaussian). The proposed active power filter is verified through a real-time digital simulator. The detailed real-time results are presented to support the feasibility of proposed control strategies.

When the supply voltages are balanced and sinusoidal, the two control strategies, instantaneous active and reactive power (p-q) and instantaneous active and reactive current (I_d-I_q), are converging to the same compensation characteristics, but when the supply voltages are distorted or unbalanced sinusoidal, these control strategies result in different degrees of compensation in harmonics. The p-q control strategy is unable to yield an adequate solution when source voltages are not ideal. Under unbalanced/nonsinusoidal conditions, the p-q control strategy does not succeed in compensating harmonic currents; notches are observed in the source current. The main reason behind the notches is that the controller failed to track the current correctly, and thereby APF fails to compensate completely. So to avoid the difficulties that occur with the p-q control strategy, we have considered the I_d-I_q control strategy.

This chapter is organized as follows: Section 1.2 deals with power quality issues, and Section 1.3 provides solutions for them. Section 1.4 introduces the importance of active power filters and solid-state devices. Classifications of active power filters are provided in Section 1.5. Section 1.6 gives details of technical and economic considerations. Selection considerations of APFs are given in Section 1.7, while Section 1.8 provides an introduction to active power filter control strategies. Finally, motivation, book objectives, and book structure are clearly outlined in Sections 1.9, 1.10, and 1.11, respectively.

1.2 Power Quality Issues

The power quality (PQ) issue is defined as any power problem manifested in voltage, current, or frequency deviations that results in damage, upset, failure, or misoperation of customer equipment. Almost all PQ issues [1–65] are closely related to power electronics in almost every aspect of commercial, domestic, and industrial application. Equipment using power electronic devices are residential appliances such as TVs and PCs, business and office equipment such as copiers and printers, and industrial equipment such as programmable logic controllers (PLCs), adjustable-speed drives (ASDs), rectifiers, inverters, and CNC (computer numerical control) tools. The PQ problem can be detected from one of the following several symptoms, depending on the type of issue involved.

Power electronics (PE) has three aspects in power distribution:

1. Power electronics introduces valuable industrial and domestic equipment.
2. Power electronics is the most important cause of harmonics: inter-harmonics, notches, and neutral currents.
3. Power electronics helps to solve PQ problems.

1.2.1 Main Causes of Poor Power Quality

The main causes of PQ issues are nonlinear loads, adjustable-speed drives, traction drives, the start of large motor loads, arc furnaces, intermittent load transients, lightning, switching transients, and faults.

1.2.2 Power Quality Problems

The PQ problems [7] are short-duration voltage variations (voltage inter-ruption, voltage sag, and voltage swell), long-duration voltage variations (undervoltage and overvoltage), voltage flicker, voltage notching, transient disturbance, and harmonic distortion. The descriptions, causes, and conse-quences of power quality issues are given in Table 1.1.

The harmonics [9] are produced by rectifiers, ASDs, soft starters, elec-tronic ballasts for discharge lamps, switched-mode power supplies, and HVAC using ASDs. Equipment affected by harmonics include transformers, motors, cables, interrupters, and capacitors (resonance). Notches are pro-duced mainly by converters, and they principally affect the electronic control devices. Neutral currents are produced by equipment using switched-mode power supplies, such as PCs, printers, photocopiers, and any triplet's gen-erator. Neutral currents seriously affect the neutral conductor temperature and transformer capability. Interharmonics are produced by static frequency converters, cycloconverters, induction motors, and arcing devices.

Equipment present different levels of sensitivity to PQ issues, depend-ing on the type of both the equipment and the disturbance. Furthermore, the effect on the PQ of electric power systems, due to the presence of PE, depends on the type of PE utilized. The maximum acceptable values of har-monic contamination are specified in Institute of Electrical and Electronics Engineers (IEEE) standards in terms of total harmonic distortion (THD).

Harmonics have frequencies that are integer multiples of the waveform's fundamental frequency. For example, given a 50 Hz fundamental waveform, the second, third, fourth, and fifth harmonic components will be at 100, 150, 200, and 250 Hz, respectively. Thus, harmonic distortion is the degree to which a waveform deviates from its pure sinusoidal values as a result of the

TABLE 1.1

Description, Causes, and Consequences of Power Quality Issues

Power Quality Issues	Description, Causes, and Consequences
Very short interruptions	**Description:** Total interruption of electrical supply for duration from a few milliseconds to less than 1 min.
	Causes: Mainly due to the opening and automatic reclosure of protection devices to decommission a faulty section of the network. The main fault causes are insulation failure, lightning, and insulator flashover.
	Consequences: Tripping of protection devices, loss of information, and malfunction of data processing equipment. Stoppage of sensitive equipment, such as ASDs, PCs, and PLCs, if they're not prepared to deal with this situation.
Voltage sag (or dip)	**Description:** Reduction in the rms voltage in the range of 10%–90% for a duration greater than half a mains cycle and less than 1 min.
	Causes: Caused by faults on the transmission or distribution network. Faults in consumer's installation. Connection of increased load demand (heavy loads) and start-up of large motors.
	Consequences: Malfunction of information technology equipment, namely, microprocessor-based control systems (PCs, PLCs, ASDs, etc.), that may lead to a process stoppage. Tripping of contactors and electromechanical relays. Disconnection and loss of efficiency in electric rotating machines.
Voltage swell	**Description:** Momentary increase of the voltage, at the power frequency, outside the normal tolerances, with duration of more than one cycle and typically less than a 1 min.
	Causes: Caused by start/stop of heavy loads, badly dimensioned power sources, badly regulated transformers (mainly during off-peak hours), system faults, load switching, and capacitor switching.
	Consequences: Data loss, flickering of lighting and screens, stoppage or damage of sensitive equipment, if the voltage values are too high.
Long interruptions	**Description:** Total interruption of electrical supply for duration of more than 1–2 min.
	Causes: Caused by equipment failure in the power system network, storms, objects (trees, cars, etc.) striking lines or poles, fire, human error, bad coordination or failure of protection devices.
	Consequences: Stoppage of all the equipment.
Undervoltage	**Description:** Decrease in the rms AC voltage to less than 90% at the power frequency for a duration longer than 1 min.
	Causes: Caused by switching on a large load or switching off a large capacitor bank.
	Consequences: The flickering of lighting and screens, giving the impression of unsteadiness of visual perception.

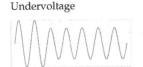

(Continued)

TABLE 1.1 (Continued)

Description, Causes, and Consequences of Power Quality Issues

Power Quality Issues	Description, Causes, and Consequences
Overvoltage 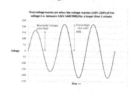	**Description:** Increase in the rms AC voltage to a level greater than 110% at the power frequency for a duration longer than 1 min. **Causes:** Caused by switching off a large load or energizing a capacitor bank. Incorrect tap settings on transformers can also cause undervoltages and overvoltages. **Consequences:** The flickering of lighting and screens, stoppage or damage of sensitive equipment, if the voltage values are too high.
Voltage unbalance 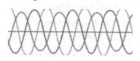	**Description:** Voltage variation in a three-phase system in which the three voltage magnitudes or phase-angle differences between them are not equal. **Causes:** Large single-phase loads (induction furnaces, traction loads), incorrect distribution of all single-phase loads by the three phases of the system (this may also be due to a fault). **Consequences:** Unbalanced systems imply the existence of a negative sequence that is harmful to all three phase loads. The most affected loads are three-phase induction machines.
Noise 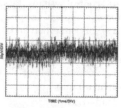	**Description:** Superimposing of high-frequency signals on the waveform of the power system frequency. **Causes:** Electromagnetic interferences provoked by Hertzian waves such as microwaves, television diffusion, and radiation due to welding machines, arc furnaces, and electronic equipment. Improper grounding may also be a cause. **Consequences:** Disturbances on sensitive electronic equipment, usually not destructive; data loss and data processing errors.
Voltage flicker 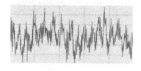	**Description:** A waveform may exhibit voltage flicker if its amplitude is modulated at frequencies less than 25 Hz, which the human eye can detect as a variation in the lamp intensity of a standard bulb. **Causes:** Caused by an arcing condition on the power system, arc furnaces, and frequent start/stop of electric motors and oscillating loads. **Consequences:** The flickering of lighting and screens, giving the impression of unsteadiness of visual perception.
Voltage notching 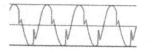	**Description:** Very fast variation of the voltage value for durations from several microseconds to a few milliseconds. These variations may reach thousands of volts, even in low voltage. It is an effect that can raise PQ issues in any facility where solid-state rectifiers are used. **Causes:** Caused by lightning, switching of lines or power factor correction capacitors, disconnection of heavy loads, the commutation of power electronic rectifiers. **Consequences:** Destruction of components and of insulation materials, data processing errors or data loss, electromagnetic interference.

(Continued)

TABLE 1.1 (Continued)

Description, Causes, and Consequences of Power Quality Issues

Power Quality Issues	Description, Causes, and Consequences
Harmonic distortion 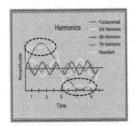	**Description:** Harmonics are periodic sinusoidal distortions of the supply voltage or load current caused by nonlinear loads. Harmonics are measured in integer multiples of the fundamental supply frequency. Voltage or current waveforms assume nonsinusoidal shape.
	Classic sources: Electric machines working above the knee of the magnetization curve (magnetic saturation), arcs (arc furnaces, fluorescent lights), welding machines, rectifiers (microprocessors, motor drives, any electronic load), and DC brush motors.
	Modern sources: All nonlinear loads, such as power electronics equipment, including ASDs, switched-mode power supplies, data processing equipment, and high-efficiency lighting.
	Consequences: Overheating of transformers, neutral overload in three-phase systems, overheating of all cables and equipment, secondary voltage distorsion of transformers, increase in power system losses, loss of efficiency in electric machines, electromagnetic interference with communication systems, failure of protective relays, explosion of capacitors, increased probability in occurrence of resonance, errors in measures when using average reading meters, nuisance tripping of thermal protections.

summation of all these harmonic elements. The ideal sine wave has zero harmonic components. In that case, there is nothing to distort this perfect wave.

Total harmonic distortion [10] of a signal is a measurement of the harmonic distortion present and is defined as the ratio of the summation of all harmonic components of the voltage or current waveform compared to the fundamental component of the voltage or current wave. The THD [11] of source current is a measure of the effective value of harmonic distortion and can be calculated as per Equations 1.1 and 1.2, in which i_1 is the root mean square (rms) value of the fundamental frequency component of current and i_n represents the rms value of the nth-order harmonic component of current as follows:

$$\text{THD} = \frac{\sqrt{\left(i_2^2 + i_3^2 + i_4^2 + \ldots + i_n^2\right)}}{i_1} \tag{1.1}$$

$$\text{THD} = \frac{\sqrt{\sum_{n=2}^{\infty} i_n^2}}{i_1} \tag{1.2}$$

1.3 Solutions for Mitigation of Power Quality Problems

There are two methodologies for the mitigation of power quality problems. The first methodology is called load conditioning, which ensures that the equipment is made less sensitive to power disturbances, allowing the operation even under significant voltage distortion. The other methodology is to install line-conditioning systems that suppress or counteract the power system disturbances.

Passive filters have been most commonly used to limit the flow of harmonic currents in distribution systems. They are usually custom-designed for the application. However, their performance is restricted to a few harmonics, and they can introduce resonance in the power system. Among the different new technical preferences available to improve power quality, active power filters have proved to be an important and flexible alternative to compensate for current and voltage disturbances in power distribution systems. The idea of active filters is relatively old, but their practical development was made possible with the new improvements in power electronics and microcomputer control strategies, as well as with cost reduction in electronic components. Active power filters [12] are becoming a viable alternative to passive filters and are quickly gaining market share as their cost becomes competitive with the passive variety. Through power electronics, the active filter introduces current or voltage components, which cancel the harmonic components of the nonlinear loads [13] or supply lines, respectively.

1.3.1 Classification of Power Filters

Classifications of power filters [7] and hybrid filters are shown in Figures 1.1 and 1.2. Power filters are classified as passive filters, active filters, or hybrid filters. Passive and active filters are categorized as series filters (passive or active), shunt filters (passive or active), or a combination of series and shunt filters (passive or active).

Hybrid filters are used in single-phase and three-phase three-wire and three-phase four-wire systems, and they are classified as series passive–shunt passive filters, series active–shunt active filters, series passive–shunt active filters, and series active–shunt passive filters

1.3.1.1 Passive Filters

Passive filters provide a low-impedance path to ground for the harmonic frequencies.

Advantages include the following:

- Harmonic reduction is possible.
- Undesirable harmonics are absent.
- Reactive power compensation is possible.

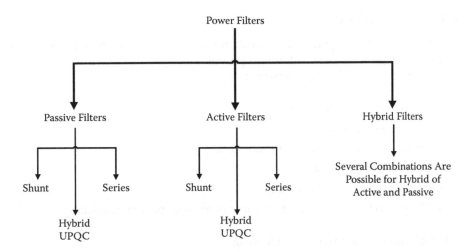

FIGURE 1.1
Classifications of power filters.

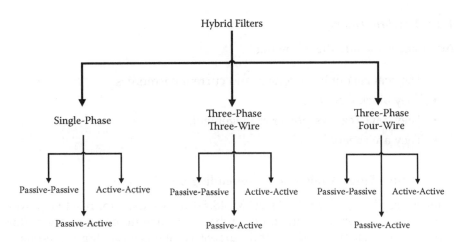

FIGURE 1.2
Classifications of hybrid filters.

Disadvantages include the following:

- Resonance with line impedance occurs.
- Tuning frequency is less accurate and requires a lot of calculations.
- They cannot be used when harmonic components vary randomly.
- They are heavy and bulky.

1.3.1.2 Active Filters

Active filters inject equal and opposite harmonics onto the power system to cancel those generated by other equipment [26].

Advantages include the following:

- They cancel out harmonics.
- They block resonance.
- They have reactive power management.
- They are small in size.
- Tuning is easy and accurate.
- They can be used when harmonic components vary randomly.

Disadvantages include the following:

- They are costly.
- There is a chance of developing inherent harmonics (because of power electronic devices).

1.3.1.3 Hybrid Filters

Advantages include the following:

- They cancel out both voltage and current harmonics.
- They block resonance.
- They have reactive power management.
- They are cheap.

1.3.2 Active Filter Applications Depending on PQ Issues

Shunt active filters [7, 9–15, 22, 24, 26, 36, 48, 50, 61] are used for reactive power compensation, voltage regulation, unbalance current compensation (for three-phase systems), and neutral current compensation (for three-phase four-wire systems). *Series active filters* [3, 28, 57] are used for reactive power compensation, voltage regulation, compensation for voltage sag and swell, and unbalance voltage compensation (for three-phase systems).

Series–shunt active filters [8, 17, 54] are used for reactive power compensation, voltage regulation, compensation for voltage sag and swell, unbalance compensation for current and voltage (for three-phase systems), and neutral current compensation (for three-phase four-wire systems). Active filter applications depending on PQ issues are given in Table 1.2.

Introduction

TABLE 1.2

Active Filter Applications Depending on PQ Issues

Active Filter Connection	Load Effect on AC Supply	AC Supply Effect on Load
Shunt active filter	• Unbalanced current compensation • Reactive power compensation • Voltage regulation • Neutral current compensation	
Series active filter	• Unbalanced voltage compensation • Reactive power compensation • Voltage regulation • Compensation for voltage sag and swell	• Voltage sag and swell • Voltage unbalance • Voltage distortion • Voltage interruption • Voltage flicker • Voltage notching
Series–shunt active filter	• Unbalanced current compensation • Unbalanced voltage compensation • Reactive power compensation • Voltage regulation • Compensation for voltage sag and swell • Neutral current compensation	• Voltage sag and swell • Voltage unbalance • Voltage distortions • Voltage interruptions • Voltage flicker • Voltage notching

1.3.3 Selection of Power Filters

The selection of power filters [7, 26, 50] is based on the following factors:

- Nature of load (voltage fed, current fed, or mixed)
- Type of supply system (single-phase and three-phase three-wire, three-phase four-wire)
- Pattern of loads (fixed, variable, fluctuating)
- Compensation required in current (harmonics, reactive power, balancing, neutral current) or voltage (harmonics, flicker, unbalance, regulation, sag, swell, spikes, notches)
- Level of compensation required (THD, individual harmonic reduction meeting specific standard, etc.)
- Environmental factors (ambient temperature, altitude, pollution, humidity, etc.)
- Cost, size, weight
- Efficiency
- Reliability

TABLE 1.3

IEEE Standard 519

Maximum Current Harmonic Distortion (% of IL)						
Individual Harmonic Order (Odd Harmonics)						
I_{ac}/I_L	<11	11 < h < 17	17 < h < 23	23 < h < 35	35 < h	TDD
<20	4.0	2.0	1.5	0.6	0.3	5.0
20–50	7.0	3.5	2.5	1.0	0.5	8.0
50–100	10.0	4.5	4.0	1.5	0.7	12.0
100–1000	12.0	5.5	5.0	2.0	1.0	15.0
>1000	15.0	7.0	6.0	2.5	1.4	20.0

Limit on Even Harmonics is 25% of Odd Harmonics limit above.
I_{ac} = Maximum short circuit current at PCC.
I_L = Maximum of fundamental component of Load Current.
TDD = Total Demand Distortion.
PCC = Point of Common Coupling.

1.3.4 IEEE Standards for Limitations of Current Harmonics

Institute of Electrical and Electronics Engineers. IEEE has stipulated certain limitations on the level of current harmonics. Table 1.3 shows the IEEE 519 harmonic current limits which specify the maximum amount of harmonic current that the customer can inject into the power system. IEEE 519 [20, 59] is developed to limit the harmonics and it is recommended to reduce the harmonic effects at any point in the entire system by establishing limits on certain harmonic indices (currents and voltages) at the point of common coupling.

1.4 Introduction to APF Technology

The growing number of power electronics-based equipment has produced a significant impact on the quality of electric power supply. Both high-power industrial loads and domestic loads cause harmonics [1] in the network voltages. At the same time, much of the equipment causing the disturbances is quite sensitive to deviations from the ideal sinusoidal line voltage. Therefore, power quality problems may originate in the system or may be caused by the consumer himself or herself. Moreover, in recent years growing concern related to power quality comes from the following:

- Consumers that are becoming gradually aware of the power quality issues [2] and more informed about the consequences of harmonics, interruptions, voltage sags, switching transients, and so forth. Motivated by deregulation, they are challenging the energy suppliers to improve the quality of the power delivered.

- The proliferation of load equipment with microprocessor-based controllers and power electronic devices that are sensitive to many types of power quality disturbances.
- Emphasis on increasing overall process productivity, which has led to the installation of high-efficiency equipment, such as adjustable-speed drives and power factor correction equipment. This in turn has resulted in an increase in harmonics injected into the power system, causing concern about their impact on the system behavior.

For an increasing number of applications, conventional equipment is proving insufficient for the mitigation of power quality problems. Harmonic distortion has traditionally been dealt with through the use of passive LC filters. However, the application of passive filters [3] for harmonic reduction may result in parallel resonances with the network impedance, overcompensation of reactive power at fundamental frequency, and poor flexibility for dynamic compensation of different frequency harmonic components.

The increased severity of power quality in power networks has attracted the attention of power engineers to develop dynamic and amendable solutions to the power quality problems.

Such equipment, generally known as active power filters [4], are also called active power line conditioners, and are able to compensate current and voltage harmonics and reactive power, regulate terminal voltage, suppress flicker, and improve voltage balance in three-phase systems. The advantage of active filtering is that it automatically adapts to changes in the network and load fluctuations. The active filters can compensate for several harmonic orders and are not affected by major changes in network characteristics, eliminating the risk of resonance between the filter and network impedance. Another advantage is that they take up very little space compared with traditional passive compensators.

The demand from electricity customers for a power supply of good quality is ever rising due to the increase of sensitive loads. This is often a stimulating task. Solid-state control of AC power using thyristors and other semiconductor switches is widely employed to feed controlled electric power to electrical loads, such as adjustable-speed drives [5], furnaces, computer power supplies, and so forth. Such controllers are also used in *high-voltage direct current* (HVDC) systems and renewable electrical power generation. As nonlinear loads, these solid-state converters draw harmonic and reactive power components of current from AC mains. In three-phase systems, they could also cause unbalance and draw excessive neutral currents. The injected harmonics, reactive power burden, unbalance, and excessive neutral currents cause low system efficiency and a poor power factor. They also cause disturbance to other consumers and interference in nearby communication networks. The increased severity of harmonic pollution in power networks has attracted the attention of power electronics and power system engineers to develop dynamic and adjustable solutions to the power quality problems. Such equipment, generally known as active power filters [1–65], are also called active power line conditioners (APLCs), instantaneous reactive power compensators (IRPCs), and active power quality conditioners (APQCs). In recent years, many publications have also appeared on harmonics [6–9, 12–16, 26, 47, 50, 60, 65], reactive power, load balancing, and neutral current compensation associated with linear and nonlinear loads [27, 32, 60].

The active power filter [1–65] technology has been under research and development for providing compensation for harmonics, reactive power, and neutral current in AC networks. APFs are also used to eliminate harmonics, regulate terminal voltage, suppress voltage flicker, and improve voltage balance in three-phase systems. This wide range of objectives is achieved either individually or in combination, depending upon the requirements and control strategy [7, 26, 36, 48, 50, 61, 63, 65] and configuration, which have to be selected appropriately.

One of the major factors in advancing the APF technology is the advent of fast self-commutating solid-state devices. In the initial stages, thyristors, bipolar junction transistors (BJTs) [4], and power MOSFETs were used for APF fabrication; later, static induction thyristors (SITs) and gate-turn-off thyristors (GTOs) were employed to develop APFs. With the introduction of insulated gate bipolar transistors (IGBTs), the APF technology got a real enhancement, and at present, they are considered as ideal solid-state devices for APFs. The improved sensor technology has also contributed to the enhanced performance of the APF. The availability of Hall effect sensors and isolation amplifiers at reasonable cost and with adequate ratings has improved APF performance.

The next breakthrough in APF development resulted from the microelectronics revolution. Starting from the use of discrete analog and digital components [5, 6], the progression has been to microprocessors, microcontrollers, and digital signal processors (DSPs) [6]. Now, it is possible to implement complex algorithms on-line for the control of the APF at reasonable cost. This development has made it possible to use different controllers, such as proportional–integral [7, 26], variable-structure, fuzzy logic [7, 11–13, 26, 50, 54, 62, 65], and neural networks [15–21], for improving the dynamic and steady-state performance of the APF. With these improvements, APFs [1–65] are capable of providing fast, corrective action, even with dynamically changing nonlinear loads.

1.5 Categorization of Active Power Filter

APFs can be classified based on converter topology, type, and number of phases. The topology can be shunt, series, or a combination of both. The converter type can be either a current source inverter (CSI) or voltage source inverter (VSI) bridge structure. The third classification is based on the number of phases, such as two-wire (single-phase) [15, 22, 27] and three- or four-wire three-phase systems [23–26, 35–41, 45–50, 61, 63, 71].

1.5.1 Converter-Based Categorization

In the development of APFs [7, 26, 50, 63] two types of converters are used: current-fed-type APF and voltage-fed-type APF. Figure 1.3 shows the current-fed pulse width modulation (PWM) inverter bridge structure. It behaves as a nonsinusoidal current source to meet the harmonic current requirement of the nonlinear load. A diode is used in series with the self-commutating device (IGBT) [29] for reverse voltage blocking. However, GTO-based configurations do not need the series diode, but they have restricted frequency of switching. They are considered sufficiently reliable but have higher losses and require higher values of parallel AC power capacitors. Moreover, they cannot be used in multilevel or multistep modes to improve performance in higher ratings.

The other converter used as an APF is a voltage-fed PWM [57] inverter structure, as shown in Figure 1.4. It has a self-supporting DC voltage bus with a large DC capacitor. It has become more dominant, since it is lighter, cheaper, and expandable to multilevel and multistep versions, to enhance the performance with lower switching frequencies. It is more popular in uninterruptible power supply (UPS)–based applications, because in the presence

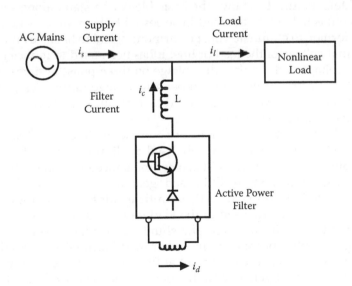

FIGURE 1.3
Current-fed-type APF.

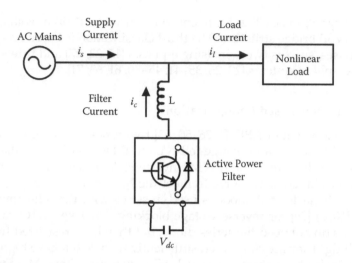

FIGURE 1.4
Voltage-fed-type APF.

of mains, the same inverter bridge can be used as an APF to eliminate harmonics of critical nonlinear loads.

1.5.2 Topology-Based Categorization

APFs can be classified based on the topology [7, 26, 50, 54] used as series-type APFs, shunt-type APFs, unified power quality conditioners (UPQCs), and hybrid filters. Figure 1.5 shows the basic block of a stand-alone series APF [5]. It is connected before the load in series with the mains, using a matching transformer, to eliminate voltage harmonics, and to balance and regulate the terminal voltage of the load or line. It has been used to reduce negative-sequence voltage and regulate the voltage on three-phase systems. It can be installed by electric utilities to compensate voltage harmonics [26] and to damp out harmonic propagation caused by resonance with line impedances and passive shunt compensators.

Figures 1.3 and 1.4 are examples of shunt active filters [5, 7, 9, 11–14, 22, 26, 36, 48, 50], which are most widely used to eliminate current harmonics, compensate reactive power, and balance unbalanced currents. The inverter circuit which is either VSI or CSI based, generates a compensating current which is injected in the line directly to mitigate for the harmonics in source current. They are mainly used at the load end, because current harmonics are injected by nonlinear loads. The shunt active filter injects equal compensating currents, opposite in phase, to cancel harmonics or reactive components of the nonlinear load current at the point of connection. It can also be used as a static var generator in the power system network for stabilizing and improving the voltage profile [9, 22].

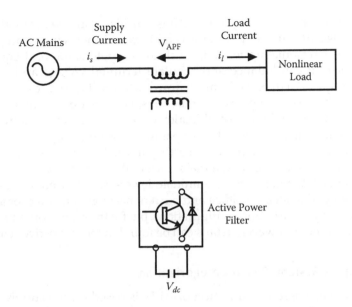

FIGURE 1.5
Series-type APF.

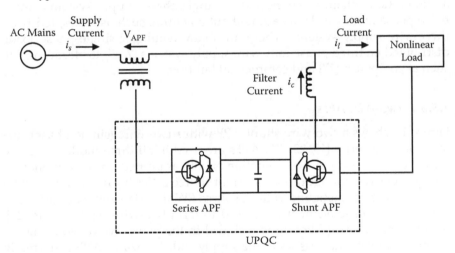

FIGURE 1.6
Unified power quality conditioner (UPQC) as universal APF.

Figure 1.6 shows a unified power quality conditioner [8, 17, 54] (also known as a universal APF), which is a combination of active series and active shunt filters. The DC link storage element (either inductor or DC bus capacitor) is shared between two current source or voltage source bridges operating as active series and active shunt compensators [17]. It is used in single-phase as well as three-phase configurations. It is considered an ideal APF that eliminates voltage and current harmonics and is capable of giving clean power to critical and harmonic-prone loads, such as computers, medical equipment, and so forth. It can balance and regulate terminal voltage and eliminate negative-sequence currents. Its main drawbacks are its large cost and control complexity because of the large number of solid-state devices involved [54].

The hybrid filter [3, 34] is a combination of series filters (either active or passive) and shunt filters (either active or passive). It is quite popular because the solid-state devices used in the active series part can be of reduced size and cost (about 5% of the load size), and a major part of the hybrid filter is made of the passive shunt L-C filter used to eliminate lower-order harmonics [34]. It has the capability of reducing voltage and current harmonics at a reasonable cost. There are many more hybrid configurations, but for the sake of brevity, they are not discussed here; however, details can be found in the respective references.

1.5.3 Supply System–Based Categorization

Supply system–based classification of APFs is based on the supply and the load system. They are single-phase two-wire APF, three-phase three-wire APF, and three-phase four-wire APFs. There are many nonlinear loads, such as domestic appliances, connected to single-phase supply systems. Some three-phase nonlinear loads are without a neutral, such as ASDs, fed from three-wire supply systems. There are many nonlinear single-phase loads distributed on three-phase four-wire supply systems [7, 25, 46–48, 50, 61, 63], such as computers [27] and commercial lighting.

1.5.3.1 Two-Wire APFs

Figure 1.7 shows a two-wire shunt APF with a two-wire (single-phase) current source inverter [15, 22, 27]. APFs are used in all three modes as active series active shunt, and a combination of both as unified line conditioners.

Both converter configurations, current source PWM bridge with inductive energy storage element and voltage source PWM bridge with capacitive DC bus energy storage elements, are used to form two-wire APF circuits [22]. In some cases, active filtering is included in the power conversion stage to improve input characteristics at the supply end. The series APF is normally used to eliminate voltage harmonics, spikes, sags, notches, and so forth, while the shunt APF is used to eliminate current harmonics and reactive power compensation [27].

1.5.3.2 Three-Wire APFs

Figure 1.8 shows a three-phase three-wire APF [23, 24, 35, 39–41, 45]. Three-phase three-wire nonlinear loads, such as ASDs, are major applications of solid-state power converters, and lately, many ASDs, and others, incorporate

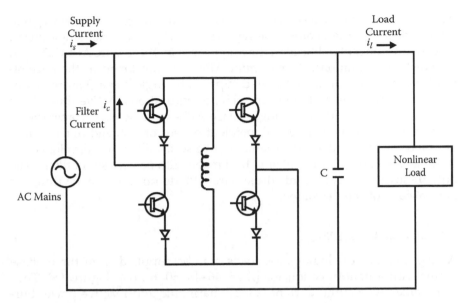

FIGURE 1.7
Two-wire shunt APF with current source inverter.

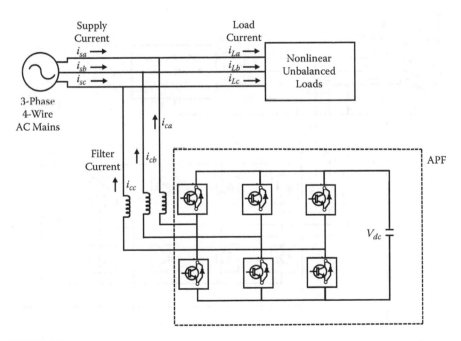

FIGURE 1.8
Three-phase three-wire APF.

APFs in their front-end design. In the three-legged configuration, each leg will develop an appropriate compensating current (i_{ca}, i_{cb}, i_{cc}) for each single phase.

A large number of publications have appeared on three-wire APFs [23] with different configurations. Active shunt APFs are developed in the current-fed type (Figure 1.3) or voltage-fed type with single-stage (Figure 1.4) or multistep/multilevel and multiseries [57] configurations. Active shunt APFs are also designed with three single-phase APFs with isolation transformers for proper voltage matching, independent phase control, and reliable compensation with unbalanced systems. Active series filters are developed for stand-alone mode (Figure 1.5) or hybrid mode with passive shunt filters. The only disadvantage is that this three-wire APF do not provide anything for neutral current compensation.

1.5.3.3 Four-Wire APFs

A large number of single-phase loads may be supplied from three-phase mains with a neutral conductor [7, 25, 46–48, 50, 61, 63] (Figure 1.9). They cause excessive neutral current [23, 24], harmonic and reactive power burden, and unbalance. To reduce these problems, four-wire APFs have been attempted. They have been developed as the active shunt mode with current-fed and voltage-fed types, active series mode, and hybrid form with active series and passive shunt mode.

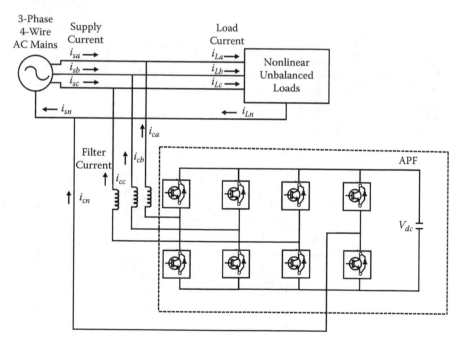

FIGURE 1.9
Three-phase four-wire four-pole shunt APF.

To employ APFs in three-phase four-wire systems, two types of configurations are possible:

1. Three-phase four-wire four-pole shunt APF
2. Three-phase four-wire capacitor midpoint shunt APF

A three-phase four-wire [7, 25, 46–48, 50, 61, 63] four-pole shunt APF is a four-leg structure, where a fourth leg is provided exclusively for neutral current compensation (Figure 1.9), and a three-phase four-wire capacitor midpoint shunt APF is a three-leg structure with the neutral conductor being connected to the midpoint of a DC link capacitor (Figure 1.10). The four-leg eight-switch APF topology is preferred for implementation, as many researchers have appointed this configuration the most proficient alternative to be used in shunt APF [26]. The three-leg six-switch split-capacitor configuration of shunt APF suffers from several shortcomings:

- The control circuit is somewhat complex.
- Voltages of the two capacitors of the split capacitor need to be properly balanced.
- Large DC link capacitors are required.

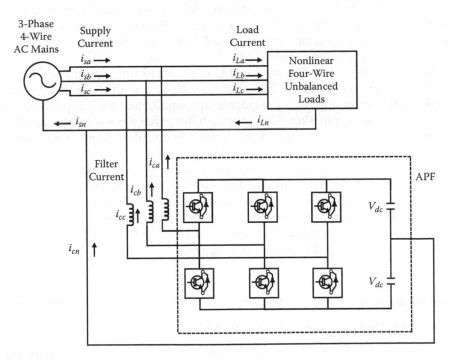

FIGURE 1.10
Three-phase four-wire capacitor midpoint shunt APF.

Despite these, three-phase four-wire capacitor midpoint [26, 50] shunt APF topology is preferred owing to its having fewer switching devices and switching losses than the eight-switch topology [4].

However, the higher-order harmonics generated in the eight-switch configuration due to frequent switching of semiconductor devices can be eliminated by the use of an RC (resistor and capacitor) high-pass filter, and switching losses occurring in the VSI can also be minimized by the use of a DC link voltage regulator.

1.6 Technical and Economic Considerations

Technical literature on the APFs has been reported since 1971 [55] and, in the last two decades, starting around 1990, has boomed. Many commercial development projects were completed [28, 56, 57] and put into practice. A number of the configurations discussed earlier have been investigated, but they have not been developed commercially because of cost and complexity considerations. Initially, reported configurations were quite general, and the rating of solid-state devices involved was substantial, which resulted in high cost. Due to these reasons, the technology could not be translated to field applications. Later on, the rating of active filtering was reduced by the introduction of supplementary passive filtering [57], without deteriorating the overall filter performance.

Moreover, modern APFs are capable of compensating quite high orders of harmonics (typically, the 25th, but nowadays it is up to 50th) dynamically. However, as high-order harmonics are small, they are compensated by using a passive ripple filter. This approach has given a boost to field applications. Another major attempt has been to separate out various compensation aspects of the APFs to reduce the size and cost. However, additional features get included on specific demand. Economic considerations were the hindrance at the initial stages of APF development [58], but now they are becoming affordable due to a reduction in the cost of the devices used. With the harmonic pollution in present-day power systems, the demand for APF is increasing. Recommended standards such as IEEE 519 [59] will result in the increased use of APFs in the coming years.

1.7 Selection Considerations of APFs

Selection of the APF for a particular application is an important task for end users and application engineers.

TABLE 1.4

Selection of APFs for Specific Application Considerations

Compensation for Specific Application	Active Series	Active Shunt	Hybrid of Active Series and Passive Shunt	Hybrid of Active Shunt and Active Series
A. Current Harmonics		**	***	*
B. Reactive Power		***	**	*
C. Load Balancing		*		
D. Neutral Current		**	*	
E. Voltage Harmonics	***		**	*
F. Voltage Regulation	***	*	**	*
G. Voltage Balancing	***		**	*
H. Voltage Flicker	**	***		*
I. Voltage Sag&Dips	***	*	**	*
J. (A + B)		***	**	*
K. (A+B+C)		**		*
L. (A+B+C+D)		*		
M. (E+F)	**			*
N. (E+F+H_I)	**			*
O. (A+E)			**	*
P. (A+B+E+F)			*	**
Q. (F+G)	**		*	
R. (B+C)		*		
S. (B+C+D)		*		
T. (A+B+G)		**	*	
U. (A+C)		*		
V. (A+D+G)		*	**	

There are widely varying application requirements, such as single-phase or three-phase three-wire and four-wire systems, requiring current- or voltage-based compensation [1–65]. Table 1.4 gives brief guidelines for the proper selection of APFs suited to the needs of individual (current-based compensation, voltage-based compensation, and voltage- and current-based compensation) requirements.

1.8 Introduction to Active Power Filter Control Strategies

For a long time one of the main concerns related to electric equipment was power factor correction, which could be done by using capacitor banks or, in some cases, reactors. For all situations, the load acted as a linear circuit drawing a sinusoidal current from a sinusoidal voltage source. Hence, the conventional power theory based on active, reactive, and apparent power definitions was sufficient for design and analysis of power systems. Nevertheless, some papers were published in the 1920s showing that the concept of reactive and apparent power loses its usefulness in nonsinusoidal conditions [66]. Then, two important approaches to power definitions under nonsinusoidal conditions were introduced by Budeanu [67, 68] in 1927 and Fryze et al. [69] in 1932. Fryze defined power in the *time domain*, whereas Budeanu did it in the *frequency domain*. At that time, nonlinear loads were negligible, and little attention was paid to this matter for a long time.

Since power electronics (PE) was introduced in the late 1960s, nonlinear loads that consume nonsinusoidal current have increased significantly. In some cases, they represent a very high percentage of the total loads. Today, it is common to find a house without linear loads, such as conventional incandescent lamps. In most cases, these lamps have been replaced by electronically controlled fluorescent lamps. In industrial applications, an induction motor that can be considered a linear load in a steady state is now equipped with a rectifier and inverter for the purpose of achieving adjustable-speed control. The induction motor, together with its drive, is no longer a linear load.

The power theories presented by Budeanu and Fryze had basic concerns related to the calculation of average power or rms values of voltage and current. The development of PE technology has brought new conditions to power theories. Exactly speaking, the new conditions have not emerged from the proliferation of power converters using power semiconductor devices such as diodes, thyristors, IGBTs, GTOs, and so on. Although these power converters have a quick response in controlling their voltages or currents, they may draw reactive power as well as harmonic current from power networks. This has made it clear that conventional power theories based on average or rms values of voltages or currents are not applicable to the analysis and design of power converters and power networks. This problem has become more serious during comprehensive analysis and design of active filters intended for reactive power compensation as well as harmonic compensation.

From the end of 1960s to the beginning of the 1970s, Erlicki and Eigeles [70], Sasaki and Machida [71], and Fukao et al. [72] published their pioneer papers presenting what can be considered a basic principle of controlled reactive power compensation. For instance, Erlicki and Eigeles [70] presented some basic ideas like "compensation of distortive power is unknown to date." They also determined that "a non-linear resistor behaves like a reactive power generator while having no energy storage elements," and presented the very first approach to active power factor control. Fukao et al. [72] stated that "by connecting a reactive power source in parallel with the load, and by controlling it in such a way as to supply reactive power to the load, the power network will only supply active power to the load. Therefore ideal power transmission would be possible."

In 1976, Gyugyi and Pelly [73] presented the idea that reactive power could be compensated by a naturally commutated cycloconverter without energy storage elements. This idea was explained from a physical point of view. However, no specific mathematical proof was presented. In 1976, Harashima et al. [74] presented, probably for the first time, the term *instantaneous reactive power* for a single phase circuit. In that same year (1976), Gyugyi and Strycula [75] used the term *active AC power filters* for the first time. A few years later, in 1981, Takahashi et al. published two papers [76, 77], giving a hint of the emergence of the instantaneous power theory, or the *p-q* theory. The *p-q* theory, in its first version, was first published in the Japanese language in 1982 [78] for a local conference, and 1983 in *Transactions of the Institute of Electrical Engineers of Japan* by Akagi et al. [79]. With a minor time lag, a paper was published in English for an international conference in 1983 [37], showing the possibility of compensating for instantaneous reactive power without energy storage elements, by Akagi et al. Then, a more complete paper, including experimental verifications, was published in the IEEE *Transactions on Industry Applications* in 1984 by Akagi et al. [39].

The *p-q theory* defines a set of instantaneous powers in the time domain. Since no restrictions are imposed on voltage or current behaviors, it is applicable to three-phase systems with or without neutral conductors, as well as to generic voltage and current waveforms. Thus, it is valid not only in steady states, but also during transient states. Contrary to other traditional power theories treating three-phase systems as three single-phase circuits, the *p-q* theory deals with all three phases at the same time, as a unity system. Therefore, this theory always considers three-phase systems together, not as a superposition or sum of three single-phase *circuits*. It was defined by using the $\alpha\beta0$ transformation, also known as the Clarke transformation [80], which consists of a real matrix that transforms three phase voltages and currents into $\alpha\beta0$ stationary reference frame currents.

In 1997, Soares et al. [44] published "Active Power Filter Control Circuit Based on the Instantaneous Active and Reactive Current I_d-I_q Method" for the IEEE Power Electronics Specialists Conference. Later in 2000, Soares et al. [43] published "An Instantaneous Active and Reactive Current Component Method for Active Filters" in *IEEE Transactions on Power Electronics*.

The I_d-I_q *control strategy* is also known as synchronous reference frame (SRF) [7, 26, 43–50]. Here, the reference frame d-q is determined by the angle θ with respect to the α-β frame used in the *p-q* theory. In the I_d-I_q control strategy [43], only the current magnitudes are transformed, and the *p-q* formulation is only performed on the instantaneous active I_d and reactive I_q components [49].

1.9 Motivation

Harmonics are mainly caused due to non-linear loads. The non-linear loads like power electronics devices and solid state devices are main sources of harmonics in the power system. These harmonics cause undesirable issues such as low power factor, overvoltage, under voltage, overheating problems, flickering, etc. Conventionally, Passive Filters are used to mitigate this problem. Passive Filters are generally connected in Shunt with the distribution system. However Passive Filters have certain limitations. They introduce resonance that effects the stability of distribution system. It has also tuning problem and not suitable for low or medium voltages.

To overcome these limitations, different configuration of Static VAR compensator (SVCs) has been proposed but it generates low harmonics and also its response time is very less for fast fluctuating loads. In recent development the APF technology is providing compensation for harmonics as well as reactive power. Current harmonics is one of the main power system problems but APF is able to mitigate it easily. The APF topology can be connected in series or Shunt or combination of both (UPQC) as well as hybrid configuration. The Shunt active power line conditioner is most commonly used than the series active power line conditioner, because most of the industrial, commercial and domestic applications need current harmonic compensation. By employing the proposed circuitry, THD is reduced to below 5% under both ideal and non-ideal conditions thereby satisfying IEEE-519 standards.

To overcome the complications encountered in power systems, APF emerged as a significant solution. The performance of APF principally depends upon the selection of the reference compensation current extraction method. Among the various APF control strategies, the instantaneous active and reactive power (p-q) method is most widely used. The p-q control strategy yields inadequate results under unbalanced or nonsinusoidal source voltage conditions. The main reason behind the notches is that the controller failed to track the current correctly, and thereby APF fails to compensate completely. So to avoid the difficulties that occur with the p-q control strategy, we have considered the I_d-I_q control strategy. By employing the I_d-I_q method, the THD in the source current after compensation can be reduced below 5% under both ideal and nonideal (unbalanced or nonsinusoidal) supply conditions, thereby satisfying IEEE 519 standards.

Even though the APF is efficient enough for load compensation, optimal performance by the APF is always desirable. The optimal harmonic compensation is possible by minimizing the undesirable losses occurring inside the APF itself. The detrimental consequences due to current harmonics, excessive neutral current, and unbalanced source current in the power system can be avoided by the use of type 1 and type 2 fuzzy logic controller (FLC)– based SHAF control strategies (p-q and I_d-I_q). Even though several controllers have been proposed, most of them yield inadequate results under nonideal supply conditions. The performance of APF with conventional PI controllers is not quite satisfactory under a range of operating conditions.

Recently, FLCs have received a great deal of attention for their applications in APFs. The advantages of fuzzy logic controllers over conventional PI controllers are that they do not require an accurate mathematical model, can work with imprecise inputs, can handle nonlinearity, and are more robust than conventional controllers. With the development of type 2 FLCs and their ability to handle uncertainty, utilizing type 2 FLCs has become very significant in recent years. The membership functions of type 2 fuzzy sets are three-dimensional and include a foot point of uncertainty, which is the new third dimension of type 2 fuzzy sets. Hence, load compensation capability of the APF can still be enhanced by using a type 2 fuzzy logic controller.

Thus, from the previous discussion it is evident that shunt APF is one of the solutions for improving the power quality by mitigating current harmonics. However, there are scopes for improving the performance of shunt APF by implementing suitable current extraction techniques and a controller for maintaining a constant DC side voltage, which is an indicator of the performance of the shunt APF. These are the primary sources of motivation for the present work.

The work presented here mainly concentrates on the use of p-q and I_d-I_q control strategies. Three types of controllers—PI controller and type 1 and type 2 fuzzy logic controllers with different fuzzy MFs (trapezoidal, triangular, and Gaussian)—have been implemented, and the results obtained with underbalanced, unbalanced, and nonsinusoidal sources are compared to evaluate the degree of compensation by the APF with different control strategies. An extensive MATLAB® simulation was carried out, and the results validate the superior functionality of type 2 FLC-based APF employing the I_d-I_q control strategy. The detailed real-time results using a real-time digital simulator are presented to support the feasibility of the proposed control strategies and controllers.

1.10 Book Objectives

From the preceding discussion, the book objectives may be outlined as follows:

- To develop shunt active filter control strategies (p-q and I_d-I_q) for extracting the three-phase reference currents and improving the power quality of power systems by mitigating the current harmonics and maintaining a constant DC link voltage.

- To develop a PI controller and the proposed type 1 and type 2 FLC-based shunt active filter control strategies with different fuzzy MFs for extracting the three-phase reference currents and maintaining a constant DC link voltage and evaluating their performance under various source-voltage conditions in a MATLAB/Simulink environment.

- To develop a hysteresis current control scheme for generation of gating signals to the devices of the APF.
- To verify the PI controller and the proposed type 1 and type 2 FLC-based shunt active filter control strategies with different fuzzy MFs with real-time digital simulator (OPAL-RT) to validate the proposed research.

1.11 Book Structure

From the introduction, it can be observed that several fundamental factors have been considered for the power quality improvement of power systems. The importance of active power filters and solid-state devices is explained in detail, and active power filter configurations and selection considerations of them are also presented in detail.

In Chapter 2, we consider PI controller–based shunt active filter control strategies (p-q and I_d-I_q). SHAF control strategies for extracting three-phase reference currents are compared, with their performance evaluated under different source voltage conditions using a PI controller. The performance of the control strategies has been evaluated in terms of harmonic mitigation and DC link voltage regulation. The detailed simulation results are presented to support the feasibility of proposed control strategies. To validate the proposed approach, the system is also implemented on real-time digital simulator hardware, and adequate results are reported for its verification.

In Chapter 3, type 1 fuzzy logic controller (FLC)–based SHAF control strategies with different fuzzy membership functions (MFs) (trapezoidal, triangular, and Gaussian) are developed for extracting three-phase reference currents, and are compared by evaluating their performance under different source voltage conditions. The performance of the control strategies has been evaluated in terms of harmonic mitigation and DC link voltage regulation. Detailed simulation and real-time results are presented to validate the proposed research.

Even though type 1 FLC-based SHAF control strategies with different fuzzy MFs are able to mitigate the harmonics, notches are presented in the source current. So to mitigate the harmonics perfectly, one has to choose a perfect controller. Therefore, in Chapter 4, type 2 FLC-based SHAF control strategies with different fuzzy MFs (trapezoidal, triangular, and Gaussian) are introduced. With this approach, the compensation capabilities of SHAF are extremely good. The detailed simulation results using MATLAB/Simulink software are presented to support the feasibility of the proposed control strategies.

In Chapter 5, a specific class of digital simulator known as a real-time simulator is introduced by answering the questions "What is real-time simulation?" "Why is it needed?" and "How does it work?" The latest trend in real-time simulation consists of exporting simulation models to a *field-programmable gate array* (FPGA). Today every researcher wants to develop his or her model in real time. The steps involved for implementation of a model from MATLAB to real time are provided in detail. The proposed type 2 FLC-based SHAF control strategies with different fuzzy MFs are verified with a real-time digital simulator (OPAL-RT) to validate the proposed research.

Last, Chapter 6 summarizes the book and looks at future work. A comparative study of PI controllers and the proposed type 1 FLC- and type 2 FLC-based SHAF control strategies with different fuzzy MFs using MATLAB and a real-time digital simulator is also presented.

2

Performance Analysis of SHAF Control Strategies Using a PI Controller

In Chapter 1, the main causes of harmonics were explained in detail. The prevalent difficulty with harmonics is voltage and current waveform distortion. Electronic equipment such as computers, battery chargers, electronic ballasts, variable-frequency drives, and switched-mode power supplies generate perilous harmonics. Harmonics issues are of great concern to engineers and building designers because they do more than distort voltage waveforms; they can overheat building wiring, cause nuisance tripping, overheat transformer units, and cause random end-user equipment failure. Thus, power quality has become more and more serious with each passing day. As a result, the active power filter (APF) has gained much more attention due to its excellent harmonic compensation. The importance of active power filters and solid-state devices is explained in detail, and APF configurations and selection considerations of them are also presented. The list of harmonics mitigation techniques is provided in Chapter 1.

In this chapter, shunt active filter control strategies (p-q and I_d-I_q) for extracting the three-phase reference currents are compared, and their performance is evaluated under different source voltage conditions (balanced, unbalanced, and nonsinusoidal) using a proportional–integral (PI) controller. The performance of the control strategies has been evaluated in terms of harmonic mitigation and DC link voltage regulation. The detailed simulation results using MATLAB®/Simulink® software are presented to support the feasibility of the proposed control strategies. To validate the proposed approach, the system is also implemented on real-time digital simulator hardware, and adequate results are reported for its verification.

This chapter is organized as follows: Section 2.1 provides details on the shunt active filter basic compensation principle. Section 2.2 outlines the classifications of shunt active filter control strategies. An introduction to DC link voltage regulation using PI controllers is provided in Section 2.3. Simulation results of p-q and I_d-I_q control strategies with PI controllers using MATLAB/Simulink are presented in Section 2.4. Real-time results of p-q and I_d-I_q control strategies with PI controllers using a real-time digital simulator are presented in Section 2.5, and finally, Section 2.6 gives concluding remarks.

2.1 Shunt Active Filter Basic Compensation Principle

Shunt active filter (SHAF) [7, 9–15, 22, 24, 26, 36, 48, 50, 61] design is an important criterion to compensate current harmonics effectively [6–9, 12–16, 26, 47, 50, 60, 65].

In brief, for perfect compensation, a controller must be capable of achieving the following requirements:

1. Extract and inject load harmonic currents
2. Maintain a constant DC link voltage
3. Avoid absorbing or generating the reactive power with fundamental frequency components.

Figure 2.1 shows a basic architecture of a three-phase four-wire shunt active filter. The active power filter [1–65] is controlled to supply the compensating current [30] to the load to cancel out the current harmonics on the AC side and reactive power flow to the source, thereby making the source current in phase with the source voltage.

Figures 2.2 and 2.3 show the basic compensation principle of the active power filter, and it serves as an energy storage element to supply the real

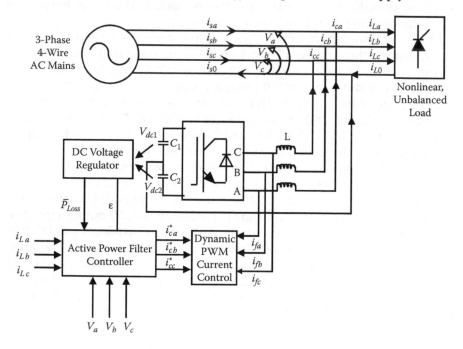

FIGURE 2.1
Basic architecture of a three-phase four-wire shunt active filter.

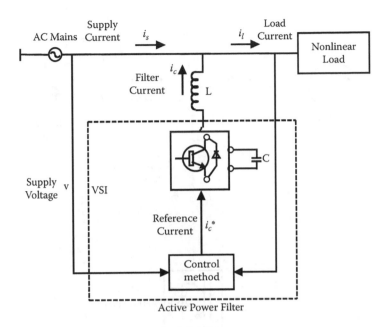

FIGURE 2.2
Basic compensation principle.

power difference between the load and source during the transient period. When the load condition changes, the real power balance between the mains and the load will be disturbed. This real power difference is to be compensated by the DC capacitor. This changes the DC capacitor voltage away from the reference voltage.

In order to keep satisfactory operation of the active filter [31], the peak value of the reference source current must be adjusted to proportionally change the real power drawn from the source. This real power absorbed/released by the capacitor compensates the real power difference between the power consumed by the load and the power supplied from the source [32]. If the DC capacitor voltage is recovered and attains the reference voltage, the real power supplied by the source is supposed to be equal to that consumed by the load again.

2.2 Active Power Filter Control Strategies

The control strategy [7] is the heart of the APF and is implemented in three stages:

1. In the first stage, the essential voltage and current signals are sensed using potential transformers (PTs) and current transformers (CTs).

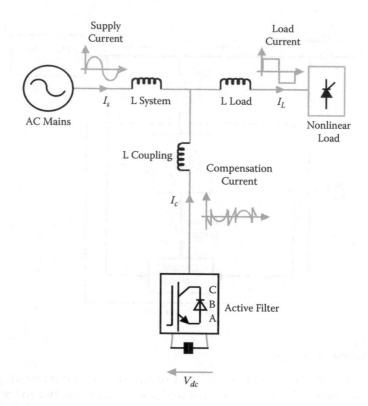

FIGURE 2.3
Compensation principle of a shunt active power filter.

2. In the second stage, compensating commands in terms of current or voltage levels are derived based on control techniques [7, 33–50] and APF configurations.

3. In the third stage of control, the gating signals for the solid-state devices of the APF are generated using pulse width modulation (PWM), hysteresis, sliding mode, PI controller, and fuzzy logic–based controllers.

The control of the APF is realized using discrete analog and digital devices or advanced microelectronic devices, such as single-chip microcomputers, digital signal processor (DSP), and so forth. There are numerous published methods that describe different topologies and different algorithms used for active filtering. In many of them, the description of a single method usually prevails, but there are publications that explain and compare a few such methods, describing their advantages and disadvantages by giving final indices as the dynamics, the total harmonic distortion (THD) reduction, the inverter efficiency, or the cost of the entire active filter.

2.2.1 Signal Conditioning

For the purpose of implementation of the control algorithm, several instantaneous voltage and current signals are required. These signals are also useful to monitor, measure, and record various performance indices, such as THD [50], power factor, active and reactive power, and so forth.

The typical voltage signals are AC terminal voltages, DC bus voltage of the APF, and voltages across series elements. The current signals to be sensed are load currents, supply currents, and compensating currents of the APF. Current signals are sensed using CTs or Hall effect current sensors. Voltage signals are sensed using either PTs or Hall effect voltage sensors or isolation amplifiers. The voltage and current signals are sometimes filtered to avoid noise problems. The filters are either hardware based (analog) or software based (digital) with low-pass, high-pass, or band pass characteristics.

2.2.2 Derivation of Compensating Signals

In an attempt to minimize the harmonic disturbances created by the non-linear loads, the choice of the active power filters improves the filtering efficiency and solves many issues existing with classical passive filters. One of the key points for a proper implementation of an active filter is to use optimized techniques for current/voltage reference generation. There exist many implementations supported by different strategies based on

- Frequency-domain harmonic detection methods
- Time-domain harmonic detection methods

The classification of harmonics methods can be done relative to the domain where the mathematical model is developed. Thus, two major directions are described here, the time-domain and frequency-domain methods [33, 34]. Such a classification is given in Table 2.1. The description of the methods will be provided in subsequent sections.

TABLE 2.1

Classification of Harmonic Detection Methods

Domain	Harmonic Detection Mehtod
Frequency-domain	Discrete Fourier Transform (DFT)
	Fast Fourier Transform (FFT)
	Recursive Discrete Fourier Transform (RDFT)
Time-domain	Instantaneous power p-q theory
	Instantaneous current (I_d-I_q) theory

Development of compensating signals in terms of either voltages or currents is the important part of APF control [35] and affects its ratings and transient as well as steady-state performance. One of the most discussed software parts (in the case of a DSP implementation) of an active filter is the harmonic detection method. In brief, it represents the part that has the capability of determining specific signal attributes (for instance, the frequency, amplitude, phase, time of occurrence, duration, energy, etc.) from an input signal (voltage, current, or both) by using a special mathematical algorithm. Then, with the achieved information, the controller is imposed to compensate for the existing distortion. It can be easily seen that if there are some errors when estimating one of the above attributes, the overall performance of the active filter could be seriously degraded in such a way that even sophisticated control algorithms cannot recover the original information. Therefore, different algorithms [7, 36] emerged for the harmonic detection, which led to a large scientific debate on which part the focus should be put on, the detection accuracy, the speed, the filter stability, easy and inexpensive implementation, and so forth.

2.2.2.1 Compensation in Frequency Domain

The control strategy in the frequency domain is based on the Fourier analysis of the distorted voltage or current signals to extract compensating commands [33, 34]. Using the Fourier transformation, the compensating harmonic components are separated from the harmonic-polluted signals and combined to generate compensating commands. The device switching frequency of the APF is kept generally more than twice the highest compensating harmonic frequency for effective compensation. The online application of the Fourier transform (solution of a set of nonlinear equations) is a cumbersome computation and results in a large response time.

2.2.2.2 Compensation in Time Domain

Control methods of the APFs in the time domain are based on instantaneous derivation of compensating commands in the form of either voltage or current signals from distorted and harmonic-polluted voltage or current signals [33–50]. There are a large number of control strategies in the time domain, which are known as

1. Instantaneous active and reactive power p-q control strategy
2. Instantaneous active and reactive current I_d-I_q control strategy

2.2.2.2.1 Instantaneous Active and Reactive Power (p-q) Control Strategy

The instantaneous active and reactive power control strategy (or p-q control strategy) was first proposed by H. Akagi and coauthors in 1984 [37], and has since been the subject of various interpretations and improvements. The instantaneous active and reactive power (*p-q*) control strategy [38, 78, 79] has been widely used and is based on Clarke's (α-β) transformation of voltage and current signals to derive compensating signals [39, 40]. In Figures 2.4 and 2.5, the entire reference current generation with the conventional *p-q* control strategy using the PI controller and fuzzy logic controller (FLC) has been illustrated. In order to maintain a constant DC link voltage [41], a PI (or FLC) branch is added to control the active power component.

The PI (or FLC) controls this small amount of active current, and then the current controller regulates this current to maintain the DC link capacitor voltage [42]. To achieve this, the DC link voltage is detected and compared with the reference voltage setting by control circuit, and then the difference is fed to the PI or FLC. According to the voltage difference, the PI (or FLC) decides how much active current is needed to maintain the DC link voltage. The output of the PI (or FLC) is an active current that is the corresponding power flow needed to maintain the DC link voltage. It is used as a part of the reference current for the current controller, which controls the inverter to provide the required compensation current.

The instantaneous active and reactive power can be computed in terms of transformed voltage and current signals. From instantaneous active and reactive powers, harmonic active and reactive powers are extracted using low-pass and high-pass filters. From harmonic active and reactive powers, using the inverse α-β transformation [39], compensating commands in terms of either currents or voltages are derived. The active filter currents are achieved from the instantaneous active and reactive powers *p* and *q* of the nonlinear load. Transformations of the phase voltages V_a, V_b, and V_c and the load currents i_{la}, i_{lb}, and i_{lc} into the α-β orthogonal coordinates using Clarke's transformation [40] are given in Equations 2.1 and 2.2 and shown in Figure 2.6.

The compensation objectives of active power filters are the harmonics present in the input currents. The present architecture represents three-phase four-wire, and it is realized with the constant power control strategy [5].

$$\begin{bmatrix} V_\alpha \\ V_\beta \\ V_0 \end{bmatrix} = C \begin{bmatrix} v_a \\ v_b \\ v_c \end{bmatrix} \; ; \; \begin{bmatrix} i_\alpha \\ i_\beta \\ i_0 \end{bmatrix} = C \begin{bmatrix} i_{la} \\ i_{lb} \\ i_{lc} \end{bmatrix} \tag{2.1}$$

$$C = \sqrt{\frac{2}{3}} \begin{bmatrix} \dfrac{1}{\sqrt{2}} & \dfrac{1}{\sqrt{2}} & \dfrac{1}{\sqrt{2}} \\ 1 & -\dfrac{1}{2} & -\dfrac{1}{2} \\ 0 & \dfrac{\sqrt{3}}{2} & -\dfrac{\sqrt{3}}{2} \end{bmatrix} \tag{2.2}$$

where *C* is the so-called transformation matrix $\|C\| = 1$ and $C^{-1} = C^T$.

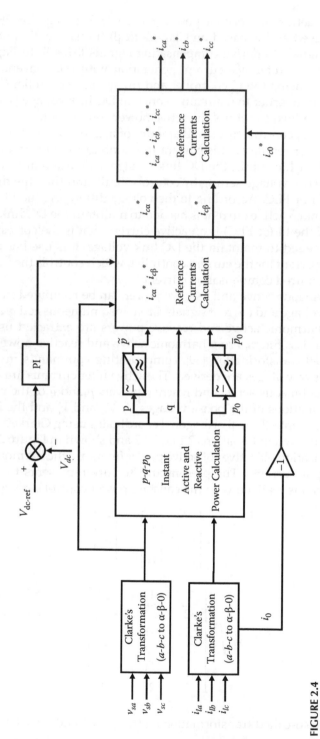

FIGURE 2.4

Reference current extraction with *p-q* control strategy using a PI controller.

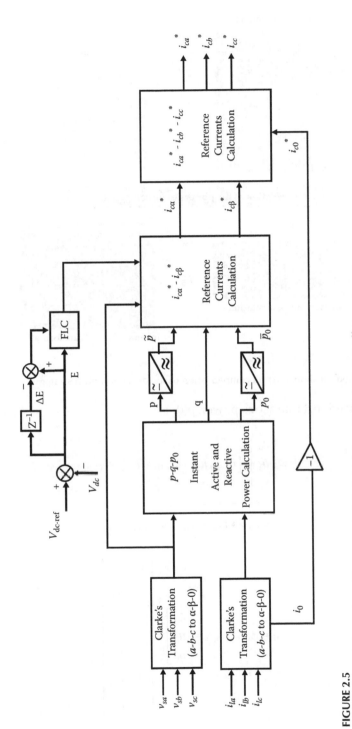

FIGURE 2.5
Reference current extraction with a *p-q* control strategy using a fuzzy logic controller.

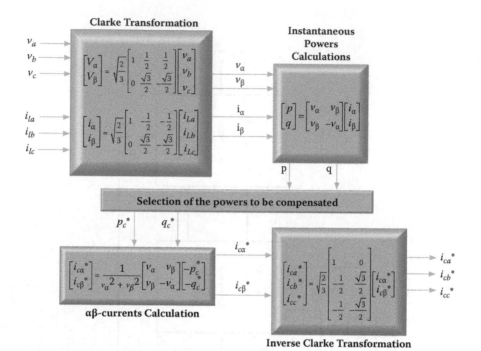

FIGURE 2.6
Control method for shunt current compensation based on a *p-q* control strategy.

Generalized instantaneous power, $p(t)$, is

$$P = \begin{bmatrix} v_\alpha \\ v_\beta \\ v_0 \end{bmatrix} \bullet \begin{bmatrix} i_{la} & i_{lb} & i_{lc} \end{bmatrix} = v_a\, i_{la} + v_b\, i_{lb} + v_c\, i_{lc} \qquad (2.3)$$

$$P = v_{\alpha\beta0} \bullet i_{\alpha\beta0} = v_\alpha\, i_\alpha + v_\beta\, i_\beta + v_0\, i_0 \qquad (2.4)$$

$$q = v_{\alpha\beta0} \times i_{\alpha\beta0} = \begin{bmatrix} q_\alpha \\ q_\beta \\ q_0 \end{bmatrix} = \begin{bmatrix} \begin{vmatrix} v_0 & v_\alpha \\ i_0 & i_\alpha \end{vmatrix} \\[12pt] \begin{vmatrix} v_\alpha & v_\beta \\ i_\alpha & i_\beta \end{vmatrix} \\[12pt] \begin{vmatrix} v_\beta & v_0 \\ i_\beta & i_0 \end{vmatrix} \end{bmatrix} \qquad (2.5)$$

$$q = \left\| \vec{q} \right\| = \sqrt{q_\alpha^2 + q_\beta^2 + q_0^2} \qquad (2.6)$$

$$\begin{bmatrix} p \\ q_\alpha \\ q_\beta \\ q_0 \end{bmatrix} = \begin{bmatrix} v_\alpha & v_\beta & v_0 \\ 0 & -v_0 & v_\beta \\ v_0 & 0 & -v_\alpha \\ -v_\beta & v_\alpha & 0 \end{bmatrix} \begin{bmatrix} i_\alpha \\ i_\beta \\ i_0 \end{bmatrix} \qquad (2.7)$$

$$\begin{bmatrix} i_\alpha \\ i_\beta \\ i_0 \end{bmatrix} = \frac{1}{v_{\alpha\beta0}^2} \begin{bmatrix} v_\alpha & 0 & v_0 & -v_\beta \\ v_\beta & -v_0 & 0 & v_\alpha \\ v_0 & v_\beta & -v_\alpha & 0 \end{bmatrix} \begin{bmatrix} p \\ q_\alpha \\ q_\beta \\ q_0 \end{bmatrix} \qquad (2.8)$$

where

$$v_{\alpha\beta0} = \begin{bmatrix} v_\alpha \\ v_\beta \\ v_0 \end{bmatrix}; \quad i_{\alpha\beta0} = \begin{bmatrix} i_\alpha \\ i_\beta \\ i_0 \end{bmatrix}; \quad v_{\alpha\beta0}^2 = v_\alpha^2 + v_\beta^2 + v_0^2$$

Instantaneous zero-sequence active current, i_{0p}, is

$$i_{0p} = \frac{v_0}{v_{\alpha\beta0}^2} P \qquad (2.9)$$

Instantaneous zero-sequence reactive current, i_{0q}, is

$$i_{0q} = \frac{v_\beta}{v_{\alpha\beta0}^2} q_\alpha - \frac{v_\alpha}{v_{\alpha\beta0}^2} q_\beta \qquad (2.10)$$

Instantaneous active current on the α-axis, $i_{\alpha p}$, is

$$i_{\alpha p} = \frac{v_\alpha}{v_{\alpha\beta0}^2} P \qquad (2.11)$$

Instantaneous reactive current on the α-axis, $i_{\alpha q}$, is

$$i_{\alpha q} = \frac{v_0}{v_{\alpha\beta0}^2} q_\beta - \frac{v_\beta}{v_{\alpha\beta0}^2} q_0 \qquad (2.12)$$

Instantaneous active current on the β-axis, $i_{\beta p}$, is

$$i_{\beta p} = \frac{v_{\beta}}{v_{\alpha\beta 0}^2} P \qquad (2.13)$$

Instantaneous reactive current on the β-axis, $i_{\beta q}$, is

$$i_{\beta q} = \frac{v_{\alpha}}{v_{\alpha\beta 0}^2} q_0 - \frac{v_0}{v_{\alpha\beta 0}^2} q_{\alpha} \qquad (2.14)$$

In the new coordinate system, the instantaneous power has two components: the zero-sequence instantaneous real power P_0 and the instantaneous
real power due to positive- and negative-sequence components $P_{\alpha\beta}$.

$$P(t) = P_0(t) + P_{\alpha\beta}(t) \qquad (2.15)$$

$$P_0(t) = v_0 i_0 \qquad (2.16)$$

$$P_{\alpha\beta}(t) = \begin{bmatrix} v_{\alpha} \\ v_{\beta} \end{bmatrix} \begin{bmatrix} i_{\alpha} & i_{\beta} \end{bmatrix} = v_{\alpha} i_{\alpha} + v_{\beta} i_{\beta} \qquad (2.17)$$

Using the above equations and considering the orthogonal nature of vec-
tors $\bar{v}$ and $\bar{q} (\bar{v} \bullet \bar{q} = 0)$, the reference source current in the αβ0 frame is

$$\begin{bmatrix} i_{s\alpha} \\ i_{s\beta} \\ i_{s0} \end{bmatrix} = \frac{1}{v_{\alpha\beta 0}^2} \begin{bmatrix} v_{\alpha} & 0 & v_0 & -v_{\beta} \\ v_{\beta} & -v_0 & 0 & v_{\alpha} \\ v_0 & v_{\beta} & -v_{\alpha} & 0 \end{bmatrix} \begin{bmatrix} p \\ q_{\alpha} \\ q_{\beta} \\ q_0 \end{bmatrix} \qquad (2.18)$$

$$\begin{bmatrix} i_{s\alpha ref} \\ i_{s\beta ref} \\ i_{s0 ref} \end{bmatrix} = \frac{1}{v_{\alpha}^2 + v_{\beta}^2} \begin{bmatrix} v_{\alpha} & 0 & v_0 & -v_{\beta} \\ v_{\beta} & -v_0 & 0 & v_{\alpha} \\ 0 & v_{\beta} & -v_{\alpha} & 0 \end{bmatrix} \begin{bmatrix} \overline{P_{L\alpha\beta}} + \overline{P_{L0}} \\ 0 \\ 0 \\ 0 \end{bmatrix} \qquad (2.19)$$

$$\begin{bmatrix} i_{s\alpha ref} \\ i_{s\beta ref} \\ i_{s0 ref} \end{bmatrix} = \frac{\overline{P_{L\alpha\beta}} + \overline{P_{L0}}}{v_{\alpha}^2 + v_{\beta}^2} \begin{bmatrix} v_{\alpha} \\ v_{\beta} \\ 0 \end{bmatrix} \qquad (2.20)$$

2.2.2.2.2 *Instantaneous Active and Reactive Current (I_d-I_q) Control Strategy*

In Figure 2.7, the entire reference current generation scheme has been illustrated. This control strategy is also known as the synchronous reference frame (SRF) [7, 26, 43–50]. Here, the reference frame d-q is determined by the angle θ with respect to the α-β frame used in the p-q theory. In the I_d-I_q control strategy [43], only the current magnitudes are transformed, and the p-q formulation is only performed on the instantaneous active I_d and reactive I_q components. In the synchronous d-q reference frame [44], voltage and current signals are transformed to a synchronously rotating frame, in which fundamental quantities become DC quantities, and then the harmonic compensating commands are extracted. The DC bus voltage feedback is generally used to achieve a self-supporting DC bus in voltage-fed APFs [45].

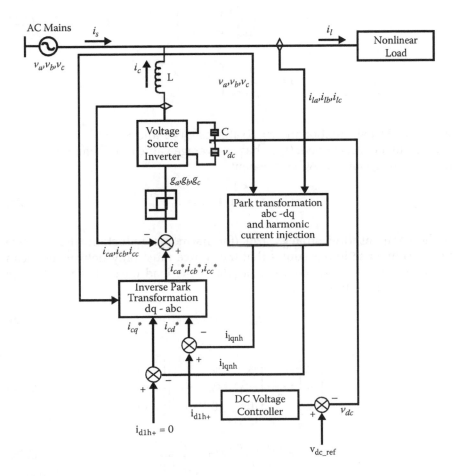

FIGURE 2.7
Control method for shunt current compensation based on the I_d-I_q control strategy.

If the d-axis has the same direction as the voltage space vector $\bar{v}$, then the zero-sequence component of the current remains invariant. Therefore, the I_d-I_q method can be expressed as follows:

$$\begin{bmatrix} i_0 \\ i_{1d} \\ i_{1q} \end{bmatrix} = \frac{1}{v_{\alpha\beta}} \begin{bmatrix} 1 & 0 & 0 \\ 0 & \cos\theta & \sin\theta \\ 0 & -\sin\theta & \cos\theta \end{bmatrix} \begin{bmatrix} i_\alpha \\ i_{1\alpha} \\ i_{1\beta} \end{bmatrix} \tag{2.21}$$

$$\begin{bmatrix} i_{ld} \\ i_{lq} \end{bmatrix} = S \begin{bmatrix} i_{l\alpha} \\ i_{l\beta} \end{bmatrix} \tag{2.22}$$

$$S = \frac{1}{v_{\alpha\beta}} \begin{bmatrix} v_\alpha & v_\beta \\ -v_\beta & v_\alpha \end{bmatrix} \tag{2.23}$$

$$S = \frac{1}{\sqrt{v_\alpha^2 + v_\beta^2}} \begin{bmatrix} v_\alpha & v_\beta \\ -v_\beta & v_\alpha \end{bmatrix} \tag{2.24}$$

where the transformation matrix S satisfies $\|S\| = 1$ and $S^{-1} = S^T$.

Each current component (I_d, I_q) has an average value or DC component and an oscillating value or AC component:

$$i_{ld} = \overline{i_{ld}} + \widetilde{i_{ld}} \quad \text{and} \quad i_{lq} = \overline{i_{lq}} + \widetilde{i_{lq}} \tag{2.25}$$

The compensating strategy [26] (for harmonic reduction and reactive power compensation) assumes that the source must only deliver the mean value of the direct-axis component [46] of the load current. The reference source current will therefore be

$$i_{sdref} = \overline{i_{Ld}}; \ i_{sqref} = i_{s0ref} = 0 \tag{2.26}$$

$$\begin{bmatrix} i_{ld} \\ i_{lq} \\ i_{l0} \end{bmatrix} = \frac{1}{v_{\alpha\beta}} \begin{bmatrix} v_\alpha & v_\beta & 0 \\ -v_\beta & v_\alpha & 0 \\ 0 & 0 & v_{\alpha\beta} \end{bmatrix} \begin{bmatrix} i_{L\alpha} \\ i_{L\beta} \\ i_{L0} \end{bmatrix} \tag{2.27}$$

$$i_{Ld} = \frac{v_\alpha i_{L\alpha} + v_\beta i_{L\beta}}{v_{\alpha\beta}} = \frac{P_{L\alpha\beta}}{\sqrt{v_\alpha^2 + v_\beta^2}} \tag{2.28}$$

The DC component of the above equation will be

$$\overline{i_{Ld}} = \left(\frac{P_{L\alpha\beta}}{v_{\alpha\beta}} \right)_{dc} = \left(\frac{P_{L\alpha\beta}}{\sqrt{v_{\alpha}^2 + v_{\beta}^2}} \right)_{dc} \qquad (2.29)$$

where the subscript *dc* means the average value of the expression within the parentheses.

Since the reference source current must be in phase with the voltage at the PCC (point of common coupling) [47] (and have no zero-sequence component), it will be calculated (in α-β-0 coordinate) by multiplying the above equation by a unit vector in the direction of the PCC voltage space vector (excluding the zero-sequence component):

$$i_{sref} = \overline{i_{Ld}} \; \frac{1}{v_{\alpha\beta}} \begin{bmatrix} v_{\alpha} \\ v_{\beta} \\ 0 \end{bmatrix} \qquad (2.30)$$

$$\begin{bmatrix} i_{s\alpha ref} \\ i_{s\beta ref} \\ i_{s0ref} \end{bmatrix} = \left(\frac{P_{L\alpha\beta}}{v_{\alpha\beta}} \right)_{dc} \frac{1}{v_{\alpha\beta}} \begin{bmatrix} v_{\alpha} \\ v_{\beta} \\ 0 \end{bmatrix} \qquad (2.31)$$

$$\begin{bmatrix} i_{s\alpha ref} \\ i_{s\beta ref} \\ i_{s0ref} \end{bmatrix} = \left(\frac{P_{L\alpha\beta}}{\sqrt{v_{\alpha}^2 + v_{\beta}^2}} \right)_{dc} \frac{1}{\sqrt{v_{\alpha}^2 + v_{\beta}^2}} \begin{bmatrix} v_{\alpha} \\ v_{\beta} \\ 0 \end{bmatrix} \qquad (2.32)$$

The reference signals thus obtained are compared with the actual compensating filter currents in a hysteresis comparator, where the actual current is forced to follow the reference and provides instantaneous compensation by the APF [48] on account of its easy implementation and quick prevail over fast current transitions. This consequently provides switching signals to trigger the IGBTs inside the inverter. Ultimately, the filter provides necessary compensation for harmonics in the source current and reactive power unbalance in the system. Figure 2.8 shows voltage and current vectors in stationary and rotating reference frames. The transformation angle θ is sensible to all voltage harmonics and unbalanced voltages; as a result, $d\theta/dt$ may not be constant.

Reference currents are extracted with the I_d-I_q control strategy using PI and FLC, which are shown in Figures 2.9 and 2.10. In order to maintain a constant DC link voltage, a PI (or FLC) branch is added to the *d*-axis in the *d-q* frame to control the active current component. The PI (or FLC) controls this small

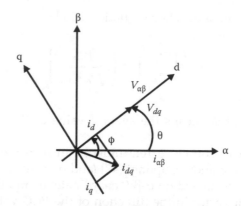

FIGURE 2.8
Instantaneous voltage and current vectors.

amount of active current, and then the current controller regulates this current to maintain the DC link capacitor voltage [50].

One of the advantages of this control strategy is that angle θ is calculated directly from main voltages, and thus makes this control strategy frequency independent by avoiding the PLL in the control circuit. Consequently, synchronizing problems with unbalanced and distorted conditions of main voltages are also eluded. Thus, I_d-I_q achieves a large-frequency operating limit, essentially by the cutoff frequency of the voltage source inverter (VSI) [49].

2.2.3 Current Control Techniques for Derivation of Gating Signals

Most applications of three-phase voltage source pulse-width-modulated (VS-PWM) converter-fed AC motor drives, active power filters, high-power-factor AC/DC converters, uninterruptible power supply (UPS) systems, and AC power supplies have a control structure comprising an internal current feedback loop. Consequently, the performance of the converter system largely depends on the quality of the applied current control strategy. Therefore, current control of PWM converters is one of the most important subjects of modern power electronics. In comparison to conventional open-loop voltage PWM converters, the current-controlled PWM (CC-PWM) [51] converters have the following advantages:

- Control of instantaneous current waveform and high accuracy
- Peak current protection
- Overload rejection
- Extremely good dynamics
- Compensation of the semiconductor voltage drop and dead times of the converter
- Compensation of the DC link and AC-side voltage changes

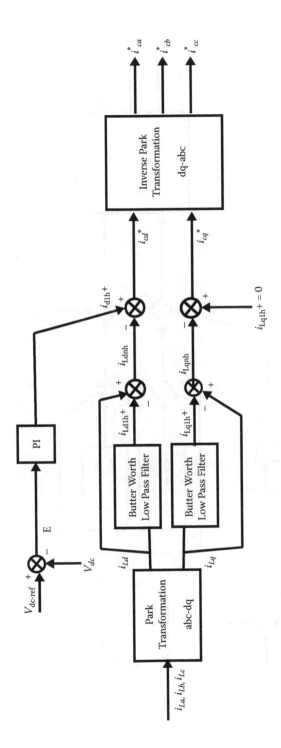

FIGURE 2.9
Reference current extraction with I_d-I_q control strategy with PI controller.

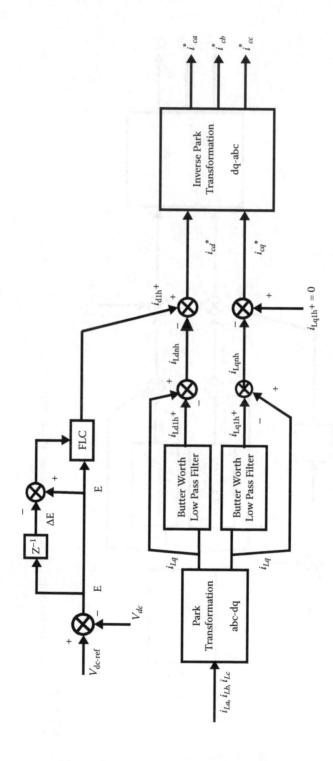

FIGURE 2.10
Reference current extraction with I_d-I_q control strategy with fuzzy logic controller.

2.2.3.1 Generation of Gating Signals to the Devices of the APF

The third stage of control of the APF is to generate gating signals for the solid-state devices of the APF based on the derived compensating commands, in terms of voltages or currents. A variety of approaches, such as hysteresis-based current control, PWM current or voltage control, deadbeat control, sliding mode of current control, PI controller, fuzzy-based current control, neural network–based current control, and so on, can be implemented, through either hardware or software, to obtain the control signals for the switching devices of the APF. Out of all these approaches, we have considered hysteresis-based current control [50–54], PI controller [7], and fuzzy logic controller [7–11, 26, 50]. In Chapter 3, the fuzzy logic controller is explained in detail.

2.2.3.1.1 Hysteresis Current Control Scheme

The hysteresis band current control technique [50] is the most suitable for the applications of current-controlled voltage source inverters in active power filters, grid-connected systems, and interior permanent magnet (IPM) machines. Basic requirements of a CC-PWM are as follows:

- No phase and amplitude errors (ideal tracking) over a wide output frequency range
- Provision of a better dynamic response of the system
- Low harmonic content
- Limited or constant switching frequency to guarantee APF operation of converter semiconductor power devices

The hysteresis band current control is characterized by unconditioned stability, very fast response, and good accuracy [50]. However, this control scheme exhibits several demerits, such as variable switching frequency, the possibility of generating resonances, and a difficult-to-design passive filter system. This unpredictable switching function affects the active filter efficiency and reliability. It may well be that fuzzy logic systems [7, 11–13, 26, 50, 54, 62, 65] resolve some of the challenges that existing methods face. The hysteresis band current control idea used for the control of active power filter line current is demonstrated in Figure 2.11a and b.

The actual source currents are monitored instantaneously, and then compared to the reference currents generated by the proposed algorithm. In order to get accurate control, switching of the IGBT [29] device should be such that the error signal should approach zero, thus providing quick response. For this reason, hysteresis current controller with a fixed band [51] that derives the switching signals of a three-phase IGBT–based VSI bridge is used.

Hysteresis control schemes are based on a nonlinear feedback loop with two-level hysteresis comparators (Figure 2.11a). The switching signals are produced directly when the error exceeds an assigned tolerance band

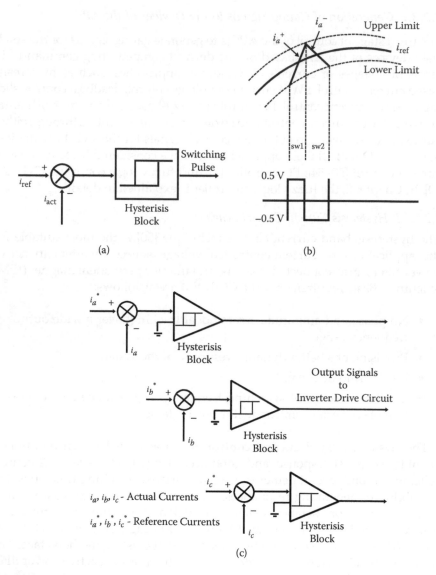

FIGURE 2.11
(a) Details of voltage and current waves with hysteresis band current controller. (a) Details of hysteresis band current controller. (c) Typical hysteresis current controller.

(Figure 2.11b). The controller generates the sinusoidal reference current of desired magnitude and frequency that is compared with the actual motor line current. If the current exceeds the upper limit of the hysteresis band [52], the upper switch of the inverter arm is turned off and the lower switch is turned on. As a result, the current starts to decay. If the current crosses the lower limit of the hysteresis band, the lower switch of the inverter arm is turned off and the upper switch is turned on. As a result, the current gets back into the hysteresis band. Hence, the actual current is forced to track the reference current within the hysteresis band [53].

The upper device and the lower device in one phase leg of VSI are switched in a complementary manner; otherwise, a dead short circuit will take place. The APF reference currents I_{sa}^*, I_{sb}^*, I_{sc}^*, compared with the sensed source currents I_{sa}, I_{sb}, I_{sc}, and the error signals are operated by the hysteresis current controller to generate the firing pulses, which activate the inverter power switches in a manner that reduces the current error. The hysteresis band current controller [54] decides the switching pattern of the active power filter. Figure 2.11c shows the hysteresis current control modulation scheme, consisting of three hysteresis comparators, one for each phase.

2.3 Introduction to DC Link Voltage Regulation

For regulating and maintaining a constant DC link capacitor voltage, the active power flowing into the active filter needs to be controlled. If the active power flowing into the filter can be controlled equal to the losses inside the filter, the DC link voltage [59] can be maintained at the desired value. The quality [60] and performance of the SHAF depend mainly on the method implemented to generate the compensating reference currents [61]. This dissertation presented two methods to get the reference current, which is the key issue in the control of the SHAF [62]. In order to maintain a constant DC link voltage and generate the compensating reference currents, we have developed the following controllers:

1. PI controller
2. Fuzzy logic controller
   a. Type 1 fuzzy logic controller with different fuzzy MFs
   b. Type 2 fuzzy logic controller with different fuzzy MFs

2.3.1 DC Link Voltage Regulation with PI Controller

Figure 2.12 shows the internal structure of the control circuit. The control scheme consists of the PI controller, limiter, and three-phase sine wave generator for reference current generation and generation of switching signals. The peak value of reference currents [63] is estimated by regulating the DC link voltage. The actual capacitor voltage is compared with a set reference

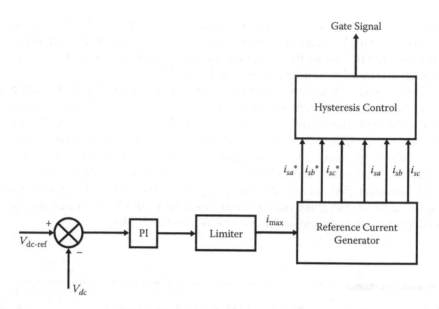

FIGURE 2.12
Conventional PI controller.

value. The error signal is then processed through a PI controller, which contributes to zero steady error in tracking the reference current signal.

The output of the PI controller [64] is considered a peak value of the supply current (I_{max}), which is composed of two components:

1. Fundamental active power component of load current
2. Loss component of APF, to maintain the average capacitor voltage to a constant value

The peak value of the current (I_{max}) so obtained is multiplied by the unit sine vectors in phase with the respective source voltages to obtain the reference compensating currents. These estimated reference currents (I_{sa}^*, I_{sb}^*, I_{sc}^*) and sensed actual currents (I_{sa}, I_{sb}, I_{sc}) are compared in a hysteresis band, which gives the error signal for the modulation technique. This error signal decides the operation of the converter switches. In this current control circuit configuration, the source/supply currents I_{sabc} are made to follow the sinusoidal reference current I_{abc}, within a fixed hysteretic band [65, 81]. The width of the hysteresis window determines the source current pattern, its harmonic spectrum, and the switching frequency of the devices.

The DC link capacitor voltage is kept constant throughout the operating range of the converter. In this scheme, each phase of the converter is controlled independently. To increase the current of a particular phase, the lower switch of the converter associated with that particular phase is turned on, while to decrease the current, the upper switch of the respective converter phase is turned on. With this, one can realize the potential and feasibility of the PI controller.

2.4 System Performance of p-q and I_d-I_q Control Strategies with PI Controller Using MATLAB/Simulink

Figures 2.13 and 2.14 highlight the performance of SHAF using the p-q and I_d-I_q control strategies with the PI controller under balanced, unbalanced, and nonsinusoidal conditions, using MATLAB/Simulink. As load is highly inductive, current draw by load is integrated with rich harmonics. Figures 2.13 and 2.14 give the details of source voltage, load current, compensation current, source current with filter, DC link voltage, and THD of the p-q and I_d-I_q control strategies with the PI controller using MATLAB under balanced, unbalanced, and nonsinusoidal supply voltage conditions.

When the supply voltages are balanced and sinusoidal, the two control strategies—instantaneous active and reactive power (p-q) control strategy and instantaneous active and reactive current (I_d-I_q) control strategy—are converging to the same compensation characteristics. But, when the supply voltages are distorted or unbalanced sinusoidal, these control strategies result in different degrees of compensation of harmonics. The p-q control strategy is unable to yield an adequate solution when source voltages are not ideal.

The THDs of the p-q control strategy with the PI controller under balanced, unbalanced, and nonsinusoidal conditions using MATLAB are 2.15%, 4.16%, and 5.31%, respectively, and the THDs of the I_d-I_q control strategy with the PI controller are 1.97%, 3.11%, and 4.93%, respectively.

2.5 System Performance of p-q and I_d-I_q Control Strategies with PI Controller Using a Real-Time Digital Simulator

In Chapter 5, a specific class of digital simulator known as a real-time simulator is introduced by answering the questions "What is real-time simulation?" "Why is it needed?" and "How does it work?" Real-time implementation of the shunt active filter model is shown in Figure 2.15. It consists of

- A host computer
- Target (real-time simulator)
- Oscilloscope

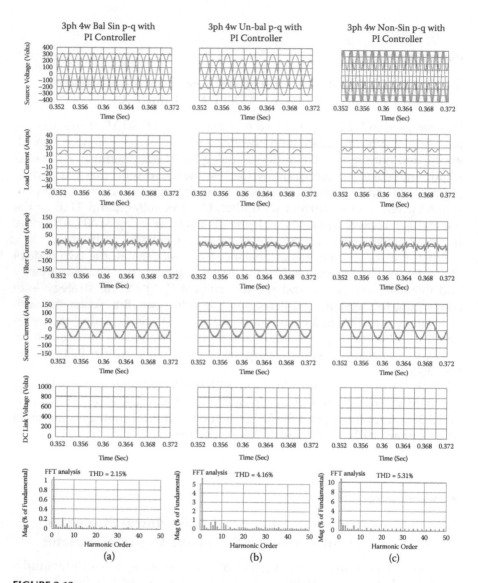

FIGURE 2.13
SHAF response using *p-q* control strategy with a PI controller using MATLAB under (a) balanced sinusoidal, (b) unbalanced sinusoidal, and (c) balanced nonsinusoidal. (i) Source voltage, (ii) load current, (iii) compensation current, (iv) source current with filter, (v) DC link voltage, and (vi) THD of source current.

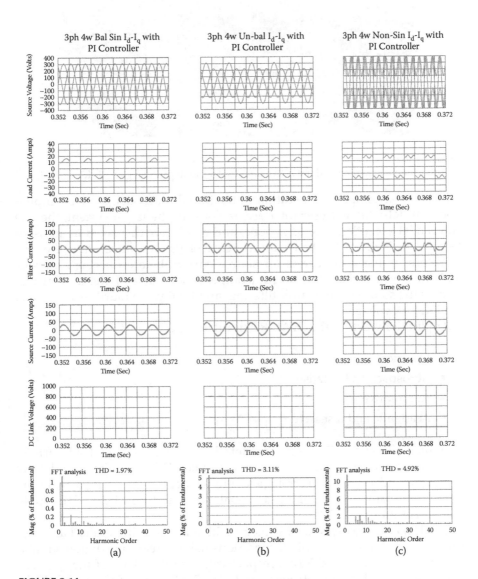

FIGURE 2.14

SHAF response using the I_d-I_q control strategy with a PI controller using MATLAB under (a) Balanced Sinusoidal, (b) Unbalanced Sinusoidal, and (c) Balanced NonSinusoidal. (i) Source voltage, (ii) load current, (iii) compensation current, (iv) source current with filter, (v) DC link voltage, and (vi) THD of source current.

FIGURE 2.15
Real-time implementation of shunt active filter control strategies using OPAL-RT.

In the host computer, an edition of the Simulink model, model compilation with RT-LAB and user interface, is done. In the target system, the I/O and model execution process is done, and results are displayed in an oscilloscope. In Figure 2.15, the oscilloscope displays four waveforms; the first waveform is the load current (or source current before filtering), the second waveform is the filter current (compensation current), the third waveform is the source current after filtering, and finally, the fourth waveform is the DC link voltage.

Application fields of the OPAL-RT simulator are

- Electrical
- Aerospace and defense
- Automotive
- Academic and research

Figures 2.16 and 2.17 present the details of source voltage, load current, compensation current, source current with filter, DC link voltage, and THD of the p-q and I_d-I_q control strategies with a PI controller using a real-time digital simulator under balanced, unbalanced, and nonsinusoidal supply voltage conditions.

The THDs of the p-q control strategy with a PI controller under balanced, unbalanced, and nonsinusoidal conditions using a real-time digital simulator are 2.21%, 4.23%, and 5.41%, respectively, and the THDs of the I_d-I_q control strategy with a PI controller are 2.04%, 3.26%, and 5.05%, respectively. Figure 2.18 clearly illustrates the THD of the source current for shunt active filter control strategies (p-q and I_d-I_q) with a PI controller using MATLAB and a real-time digital simulator under various source voltage conditions.

While considering the p-q control strategy with the PI controller, SHAF does not succeed in compensating harmonic currents; notches are observed

3ph 4w Bal Sin p-q with PI Controller
(RT DS Hardware)

3ph 4w Un-bal p-q with PI Controller
(RT DS Hardware)

3ph 4w Non-Sin p-q with PI Controller
(RT DS Hardware)

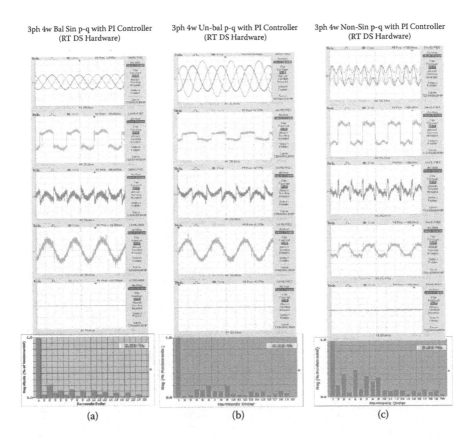

(a)   (b)   (c)

FIGURE 2.16
SHAF response using *p-q* control strategy with PI controller using real-time digital simulator under (a) balanced sinusoidal, (b) unbalanced sinusoidal, and (c) balanced nonsinusoidal. (i) Source voltage, (ii) load current (scale 20 A/div), (iii) compensation current (scale 20
A/div), (iv) source current (scale 30 A/div) with filter, (v) DC link voltage, and (vi) THD of
source current.

in the source current. The main reason behind the notches is that the controller failed to track the current correctly, and thereby APF fails to compensate
completely. The PI controller is unable to mitigate the harmonics effectively,
and notches are observed in the source current. So to mitigate the harmonics
perfectly, one has to choose the perfect controller. So to avoid the difficulties
that occur with the PI controller, we have considered the type 1 fuzzy logic
controller (type 1 FLC)–based *p-q* and I_d-I_q control strategies with different
fuzzy MFs (trapezoidal, triangular, and Gaussian MF). The system parameters are given in Table 2.2.

3ph 4w Bal Sin I_d-I_q with PI Controller   3ph 4w Un-bal I_d-I_q with PI Controller 3ph 4w Bal Non-Sin I_d-I_q with PI Controller
(RT DS Hardware)   (RT DS Hardware)   (RT DS Hardware)

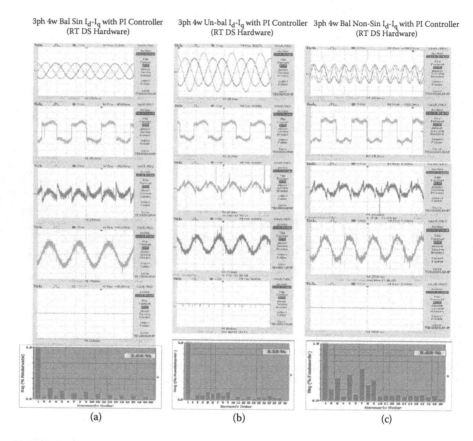

(a)   (b)   (c)

FIGURE 2.17

SHAF response using I_d-I_q control strategy with PI controller using real-time digital simulator under (a) balanced sinusoidal, (b) unbalanced sinusoidal, and (c) balanced nonsinusoidal. (i) Source voltage, (ii) load current (scale 20 A/div), (iii) compensation current (scale 20 A/div), (iv) source current (scale 30 A/div) with filter, (v) DC link voltage, and (vi) THD of source current.

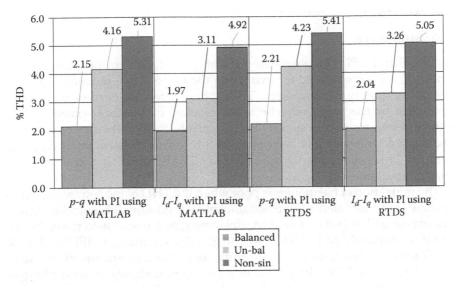

FIGURE 2.18
THD of source current for p-q and I_d-I_q control strategies with PI controller using MATLAB and real-time digital simulator.

TABLE 2.2

System Parameters

Parameter	Value
Supply voltage	$V_s = 311.12$ V
Source resistance	$R_s = 0.1\ \Omega$
Source inductance	$L_s = 1$ mH
Filter phase-branch-resistance	$R_f = 0.01\ \Omega$
Filter phase-branch inductance	$L_f = 0.1$ mH
DC link capacitance	$C_{dc} = 3000\ \mu f$
DC link voltage	$V_{dc} = 800$ V
Hysterisis band	± 0.2 A
Load	Diode rectifier
	Snubber resistance $R_{sn} = 500\ \Omega$
	Snubber capacitance $L_{sn} = 250$ e-9 f
Load resistance	$R_L = 15\ \Omega$
Load inductance	$L_L = 6$ mH

2.6 Summary

In this chapter, PI controller–based shunt active filter control strategies (p-q and I_d-I_q) were considered for extracting the three-phase reference currents for power quality improvement. Three-phase reference current waveforms generated by the proposed scheme are tracked by the three-phase voltage source converter in a hysteresis band control scheme. The performance of the control strategies has been evaluated in terms of harmonic mitigation and DC link voltage regulation using MATLAB/Simulink and a real-time digital simulator.

The mitigation of harmonics is poor when the THD of the source current is greater. Under unbalanced and nonsinusoidal conditions, PI controller–based shunt active filter (p-q and I_d-I_q) control strategies are unable to mitigate harmonics completely and THD is close to 5%. But, according to IEEE 519–1992, THD must be less than 5%. So to mitigate harmonics effectively, we have considered type 1 FLC-based p-q and I_d-I_q control strategies with different fuzzy MFs (trapezoidal, triangular, and Gaussian MF).

3

Performance Analysis of SHAF Control Strategies Using Type 1 FLC with Different Fuzzy MFs

In Chapter 2, shunt active filter (SHAF) control strategies were discussed. It is concluded that under unbalanced and nonsinusoidal conditions, the PI controller is unable to mitigate the harmonics completely and THD is close to 5%. But, according to IEEE 519–1992, THD must be less than 5%. So to mitigate the harmonics perfectly, one has to choose the perfect controller.

In this chapter, we have developed type 1 fuzzy logic controller (FLC) with different fuzzy membership function (MFs) (trapezoidal, triangular, and Gaussian). The shunt active filter control strategies (p-q and I_d-I_q) for extracting the three-phase reference currents are compared, evaluating their performance under different source voltage conditions using type 1 FLC. The performance of the control strategies has been evaluated in terms of harmonic mitigation and DC link voltage regulation. The detailed simulation results using MATLAB®/Simulink® software are presented to support the feasibility of the proposed control strategies. To validate the proposed approach, the system is also implemented on a real-time digital simulator, and adequate results are reported for its verification.

This chapter is organized as follows: Section 3.1 provides the details of the type 1 fuzzy logic controller, which include types of fuzzy inference systems, defuzzification methods, design of control rules, and rule base. Simulation results of type 1 FLC-based p-q and I_d-I_q control strategies with different fuzzy MFs using MATLAB are presented in Sections 3.2 and 3.3, respectively. Real-time results of type 1 FLC-based p-q and I_d-I_q control strategies with different fuzzy MFs using a real-time digital simulator are presented in Sections 3.4 and 3.5, respectively. Section 3.6 provides the comparative study, and finally, Section 3.7 gives concluding remarks.

3.1 Type 1 Fuzzy Logic Controller

The concept of fuzzy logic (FL) [82] was proposed by Professor Lofti Zadeh in 1965, at first as a way of processing data by allowing partial set membership

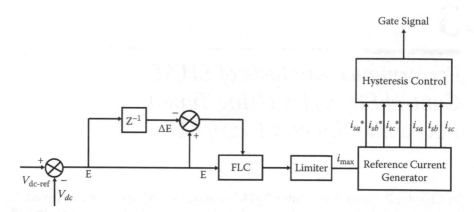

FIGURE 3.1
Proposed fuzzy logic controller.

rather than crisp membership. Soon after [83–107], it was proved to be an excellent choice for many control system applications [84] since it mimics human control logic.

Figure 3.1 shows the internal structure of the control circuit. The control scheme consists of the fuzzy logic controller (FLC) [85], limiter, and three-phase sine wave generator for reference current generation and generation of signals for the switching devices of the active power filter (APF). The peak value of reference currents is estimated by regulating the DC link voltage [26]. The actual capacitor voltage is compared with a set reference value. The error signal is then processed through FLC, which contributes to zero steady error in tracking the reference current signal.

The block diagram [50] of the fuzzy logic controller is shown in Figure 3.2. It consists of the following blocks:

- Fuzzification
- Fuzzy inference system

- Knowledge base
- Defuzzification

3.1.1 Fuzzification

The process of converting a numerical variable (real number) to a linguistic variable (fuzzy number) is called fuzzification [86].

3.1.2 Fuzzy Inference Systems

The fuzzy inference system (FIS) [87] is a popular computing framework based on the concepts of fuzzy set theory [88], fuzzy if–then rules, and fuzzy reasoning. It has found successful applications in a wide variety of fields, such as expert systems, data classification, decision analysis, time-series prediction, automatic control, pattern recognition, and robotics. Because of its

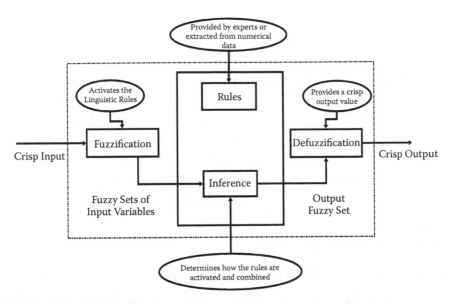

FIGURE 3.2
Block diagram of fuzzy logic controller (FLC).

multidisciplinary nature, the FIS is known by various other names, such as fuzzy rule-based system, fuzzy expert system [89], fuzzy model [90], fuzzy logic controller [91], fuzzy associative memory [92], and simply fuzzy system.

The basic structure of a FIS consists of three components: a *rule base*, which contains a selection of rules; a *database*, which defines the membership functions used in the fuzzy rules; and a *reasoning mechanism*, which performs the inference procedure upon the rules and given facts to derive a reasonable output. The basic FIS [93] can take either crisp inputs or fuzzy singletons, but the outputs it produces are almost always fuzzy sets. Sometimes it is necessary to have a crisp output, especially in a situation where a FIS is used as a controller. Therefore, we need a method of defuzzification to extract a crisp value that best represents a fuzzy set. A FIS with a crisp output is shown in Figure 3.3, where the dashed line indicates a basic fuzzy inference system with fuzzy output, and a defuzzification [94] block serves the purpose of transforming an output fuzzy set into a crisp value. Defuzzification strategies [95] are explained in Section 3.1.3.

The Mamdani fuzzy implication models [97] are used for evaluation of individual rules. There are mainly two types of implication methods used. The differences between these two fuzzy inference systems lie in the consequences of their fuzzy rules, and thus their aggregation, and differ accordingly.

1. Mamdani max-min composition scheme
2. Mamdani max-prod composition scheme

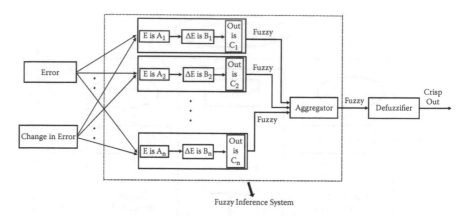

Fuzzy Inference System

FIGURE 3.3
Block diagram of fuzzy inference system.

3.1.2.1 Mamdani Max-Min Composition Scheme

The Mamdani fuzzy inference system using the max-min composition [50] scheme is shown in Figure 3.4. In this figure, the aggregation used is maximum operation and the implication is minimum operation.

3.1.2.2 Mamdani Max-Prod Composition Scheme

The Mamdani fuzzy inference system using a max-prod composition [96] scheme is shown in Figure 3.5. In this figure, the aggregation used is maximum operation and the implication is product operation.

3.1.3 Defuzzification

The rules of FLC generate required output in a linguistic variable [97] (fuzzy number), and according to real-world requirements, linguistic variables have to be transformed to crisp output (real number).

$$\mu_A(x) = \text{defuzz}(x, mf, \text{type}) \tag{3.1}$$

where defuzz(x, mf, type) gives a defuzzified value of a membership function (MF) positioned at an associated variable value x using one of several defuzzification methods [98] according to the argument type. The variable type can be one of the following:

- COA: Centroid of area
- BOA: Bisector of area
- MOM: Mean value of maximum
- SOM: Smallest (absolute) value of maximum
- LOM: Largest (absolute) value of maximum

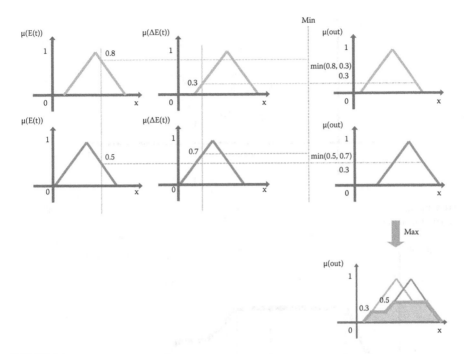

FIGURE 3.4
The Mamdani fuzzy inference system using max-min composition.

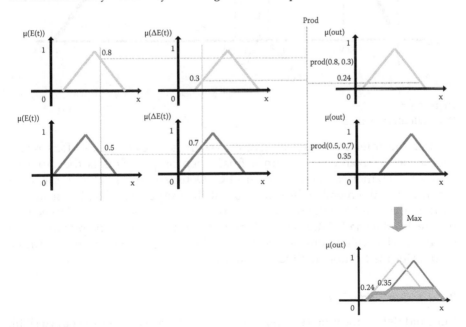

FIGURE 3.5
The Mamdani fuzzy inference system using max-prod composition.

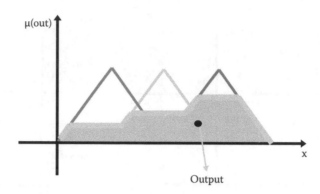

FIGURE 3.6
Defuzzification.

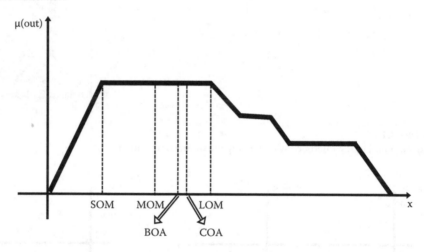

FIGURE 3.7
Defuzzification methods.

After the inference step, the overall result is a fuzzy value. This result should be defuzzified to obtain a final crisp output. This is the purpose of the defuzzifier component of an FLC. Defuzzification [99] is performed according to the membership function of the output variable. For instance, assume that we have the result in Figure 3.4 at the end of the inference. In Figure 3.6, the shaded areas all belong to the fuzzy result. The purpose is to obtain a crisp value, represented with a dot in Figure 3.6, from this fuzzy result. Defuzzification methods are shown in Figure 3.7.

3.1.3.1 *Centroid of Area*

Centroid defuzzification returns the center of the area under the curve [100]. Mathematically, a centroid of area (COA) can be expressed as

$$COA = \frac{\int_a^b \mu_A(x)\, x\, dx}{\int_a^b \mu_A(x)\, dx} \qquad (3.2)$$

With a discretized universe of discourse, the expression is

$$COA = \frac{\sum_{i=1}^{n} \mu_A(x_i)\, x_i}{\sum_{i=1}^{n} \mu_A(x_i)} \qquad (3.3)$$

3.1.3.2 Bisector of Area

The bisector is the vertical line that will divide the region into two sub-regions of equal area. It is sometimes, but not always, coincident with the centroid line. Mathematically, BOA can be expressed as

$$\int_{\alpha}^{X_{BOA}} \mu_A(x)dx = \int_{X_{BOA}}^{\beta} \mu_A(x)dx \qquad (3.4)$$

3.1.3.3 Mean, Smallest, and Largest of Maximum

MOM, SOM, and LOM stand for mean, smallest, and largest of maximum, respectively. These three methods key off the maximum value assumed by the aggregate membership function. If the aggregate membership function has a unique maximum, then MOM, SOM, and LOM [101] all take on the same value. Mathematically, MOM can be expressed as

$$X_{MOM} = \frac{\int_{X'} x\, dx}{\int_{X'} dx} \qquad (3.5)$$

where

$$X' = \left\{ x; \mu_A(x) = \mu^* \right\}$$

The fuzzy controller is characterized as follows:

- Seven fuzzy sets for each input and output
- Fuzzification using a continuous universe of discourse
- Implication using Mamdani's min operator
- Defuzzification using the centroid method

3.1.4 Design of Control Rules

In an FLC [102], a rule base is constructed to control the output variable. A fuzzy rule is a simple if–then rule with a condition and a conclusion. The fuzzy control rule design involves defining rules that relate the input variables to the output model properties. As FLC is independent of the system model, the design is mainly based on the intuitive feeling for and experience of the process [103]. The rules are expressed in English like language with syntax, such as if {error E is X and change in error ΔE is Y}, then {control output is Z}.

For better control performance, finer fuzzy petitioned subspaces {NB (negative big), NM (negative medium), NS (negative small), ZE (zero), PS (positive small), PM (positive medium), and PB (positive big)} are used. These seven membership functions are the same for input and output and characterized using trapezoidal, triangular, and Gaussian membership functions [104], as can be seen in Figures 3.8 through 3.10.

Fuzzy sets [105, 106] are chosen based on the error in the DC link voltage. We have considered a 7×7 membership function. Readers may raise questions such as the following: Why a 7×7 MF only? Why not a 3×3 or 5×5 MF? While considering a 3×3 or 5×5 MF, the SHAF is unable to maintain a constant DC link voltage, and the error voltage develops due to the voltage difference. Because of this error voltage, the system is unable to track the currents perfectly, and THD also becomes high. So to avoid these difficulties, we have considered 7×7 membership functions. With the use of 7×7 MF, SHAF is able to maintain a constant DC link voltage, which is nearly equal to the reference voltage, and it also tracks the currents perfectly [50].

3.1.4.1 Trapezoidal Membership Function

The trapezoidal curve is a function of a vector, x, and depends on four scalar parameters [50, 65, 98, 104], a, b, c, and d, as given by

$$\mu_A(x) = \begin{cases} 0 & \text{if } x \leq a \\ \dfrac{x-a}{b-a} & \text{if } a \leq x \leq b \\ 1 & \text{if } b \leq x \leq c \\ \dfrac{d-x}{d-c} & \text{if } c \leq x \leq d \\ 0 & \text{if } x \geq d \end{cases} \tag{3.6}$$

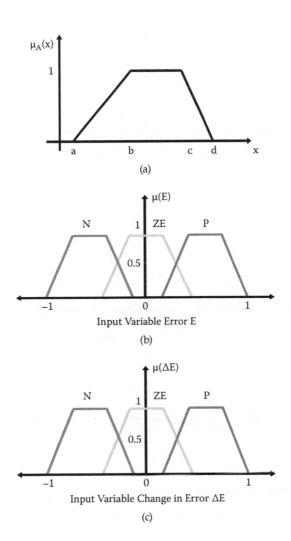

FIGURE 3.8
(a) Trapezoidal MF. (b) Input variable error E trapezoidal MF 3×3. (c) Input change in error normalized trapezoidal MF 3×3. (d) Output I_{max} normalized trapezoidal MF 3×3. (e) Input variable error E trapezoidal MF 5×5. (f) Input change in error normalized trapezoidal MF 5×5. (g) Output I_{max} normalized trapezoidal MF 5×5. (h) Input variable error E trapezoidal MF 7×7. (i) Input change in error normalized trapezoidal MF 7×7. (j) Output I_{max} normalized trapezoidal MF 7×7. *(Continued)*

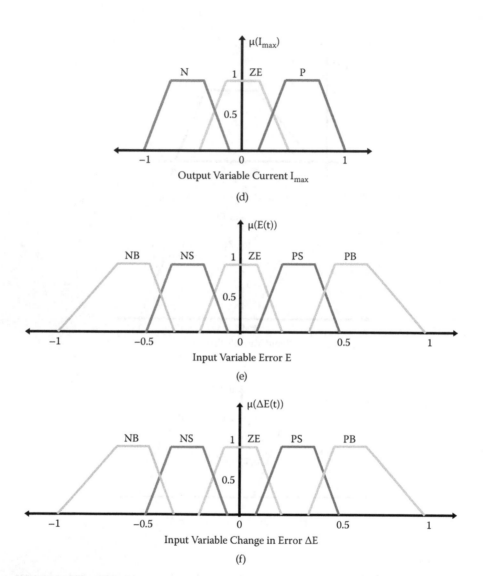

FIGURE 3.8 (Continued)
(a) Trapezoidal MF. (b) Input variable error E trapezoidal MF 3×3. (c) Input change in error normalized trapezoidal MF 3×3. (d) Output I_{max} normalized trapezoidal MF 3×3. (e) Input variable error E trapezoidal MF 5×5. (f) Input change in error normalized trapezoidal MF 5×5. (g) Output I_{max} normalized trapezoidal MF 5×5. (h) Input variable error E trapezoidal MF 7×7. (i) Input change in error normalized trapezoidal MF 7×7. (j) Output I_{max} normalized trapezoidal MF 7×7.

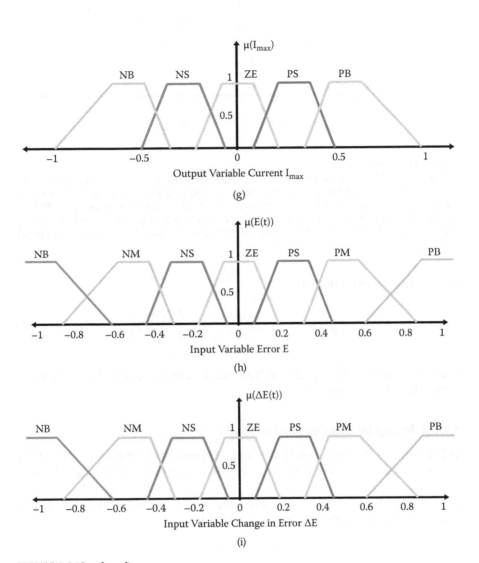

FIGURE 3.8 (Continued)
(a) Trapezoidal MF. (b) Input variable error E trapezoidal MF 3×3. (c) Input change in error normalized trapezoidal MF 3×3. (d) Output I_{max} normalized trapezoidal MF 3×3. (e) Input variable error E trapezoidal MF 5×5. (f) Input change in error normalized trapezoidal MF 5×5. (g) Output I_{max} normalized trapezoidal MF 5×5. (h) Input variable error E trapezoidal MF 7×7. (i) Input change in error normalized trapezoidal MF 7×7. (j) Output I_{max} normalized trapezoidal MF 7×7.

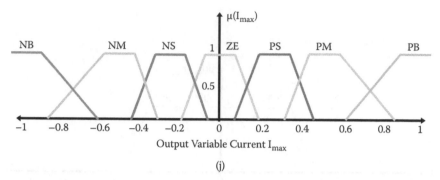

(j)

FIGURE 3.8 (Continued)
(a) Trapezoidal MF. (b) Input variable error E trapezoidal MF 3×3. (c) Input change in error normalized trapezoidal MF 3×3. (d) Output I_{max} normalized trapezoidal MF 3×3. (e) Input variable error E trapezoidal MF 5×5. (f) Input change in error normalized trapezoidal MF 5×5. (g) Output I_{max} normalized trapezoidal MF 5×5. (h) Input variable error E trapezoidal MF 7×7. (i) Input change in error normalized trapezoidal MF 7×7. (j) Output I_{max} normalized trapezoidal MF 7×7.

It can also be represented as

$$\mu_A(x) = \max\left(\min\left(\frac{x-a}{b-a}, 1, \frac{d-x}{d-c} \right), 0 \right) \tag{3.7}$$

The parameters a and d locate the "feet" of the trapezoid, and the parameters b and c locate the "shoulders."

3.1.4.2 Triangular Membership Function

In Figure 3.9a, the parameters a, b, and c represent the x coordinates of the three vertices of $\mu_A(x)$ in a fuzzy set A [50] (a, lower boundary; c, upper boundary where membership degree is zero; and b, the center where membership degree is 1). The triangular curve is a function of a vector, x, and depends on three scalar parameters, a, b, and c, as given by

$$\mu_A(x) = \begin{cases} 0 & \text{if } x \leq a \\ \dfrac{x-a}{b-a} & \text{if } a \leq x \leq b \\ \dfrac{c-x}{c-b} & \text{if } b \leq x \leq c \\ 0 & \text{if } x \geq c \end{cases} \tag{3.8}$$

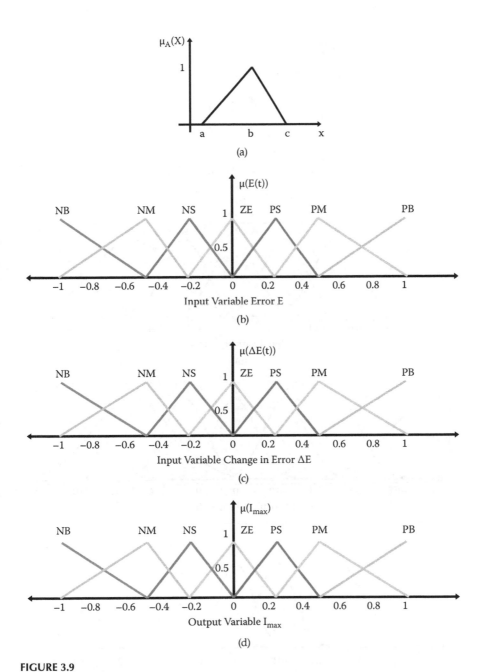

FIGURE 3.9
(a) Triangular MF. (b) Input variable error E triangular MF 7×7. (c) Input change in error normalized triangular MF 7×7. (d) Output I_{max} normalized triangular MF 7×7.

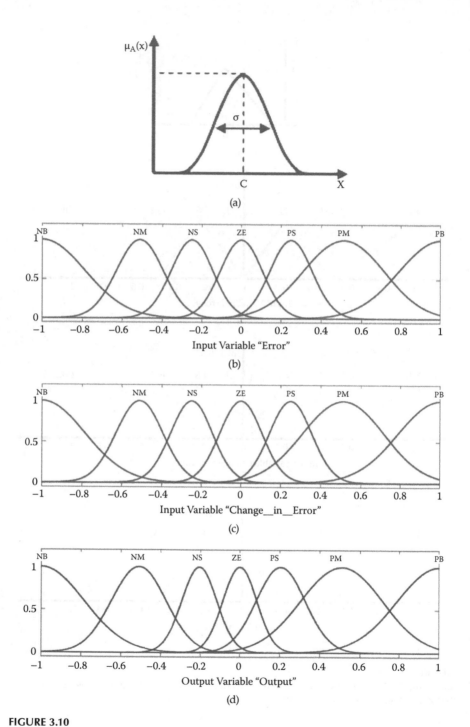

FIGURE 3.10
(a) Gaussian MF. (b) Input variable error E Gaussian MF 7×7. (c) Input change in error normalized Gaussian MF 7×7. (d) Output I_{max} normalized Gaussian MF 7×7.

It can also be represented as

$$\mu_A(x) = \max\left(\min\left(\frac{x-a}{b-a}, \frac{c-x}{c-b} \right), 0 \right)$$ (3.9)

3.1.4.3 Gaussian Membership Function

The Gaussian [26, 50, 65, 98, 104] curve is a function of a vector, x, and depends on three scalar parameters c, s, and m, as given by

$$\mu_A(x,c,s,m) = \exp\left[-\frac{1}{2}\left| \frac{x-c}{\sigma} \right|^m \right]$$ (3.10)

where c is the center, σ is the width, and m is the fuzzification factor.

3.1.5 Rule Base

The elements of this rule base [7, 26, 50, 65, 98, 104] table are determined based on the theory that in the transient state, large errors need coarse control, which requires coarse input/output variables. In the steady state, small errors need fine control, which requires fine input/output variables [104]. Based on this, the elements of the rule table are obtained as shown in Tables 3.1 through 3.3 with E and ρE as inputs.

Figure 3.11 shows the fuzzy inference system. It consists of

- Fuzzy inference system (FIS) editor
- Membership function editor
- Rule editor
- Rule viewer
- Surface viewer

TABLE 3.1

Rule Base 3 × 3

E \ ΔE	N	Z	P
N	N	N	z
z	N	z	P
P	z	P	P

TABLE 3.2

Rule Base 5 × 5

E \ ΔE	NB	NS	z	PS	PB
NB	NB	NB	NB	NS	Z
NS	NB	NB	NS	Z	PS
Z	NB	NS	Z	PS	PB
PS	NS	Z	PS	PB	PB
PB	Z	PS	PB	PB	PB

TABLE 3.3

Rule Base 7 × 7

E \ ΔE	NB	NM	NS	z	PS	PM	PB
NB	NB	NB	NB	NB	NM	NS	z
NM	NB	NB	NB	NM	NS	Z	PS
NS	NB	NB	NM	NS	Z	PS	PM
z	NB	NM	NS	Z	PS	PM	PB
PS	NM	NS	Z	PS	PM	PB	PB
PM	NS	Z	PS	PM	PB	PB	PB
PB	Z	PS	PM	PB	PB	PB	PB

The FIS editor [104] handles the high-level issues for the system: How many input and output variables? What are their names? The present model consists of two inputs and one output: error, change in error, and output, respectively. The membership function editor is used to define the shapes of all the membership functions associated with each variable. The rule editor is used for editing the list of rules that defines the behavior of the system. In the present model, 49 rules are developed.

The rule viewer and the surface viewer are used for looking at, as opposed to editing, the FIS. They are strictly read-only tools. Used as a diagnostic, they can show which rules are active or how individual membership function [26] shapes are influencing the results. The surface viewer is used to display the dependency of one of the outputs on any one or two of the inputs; that is, it generates and plots an output surface map for the system.

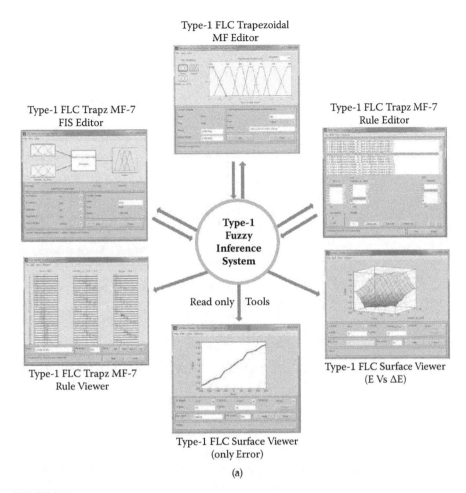

Type-1 FLC Trapezoidal
MF Editor

Type-1 FLC Trapz MF-7
FIS Editor

Type-1 FLC Trapz MF-7
Rule Editor

Type-1
Fuzzy
Inference
System

Read only Tools

Type-1 FLC Trapz MF-7
Rule Viewer

Type-1 FLC Surface Viewer
(E Vs ΔE)

Type-1 FLC Surface Viewer
(only Error)

(a)

FIGURE 3.11
(a) Type 1 fuzzy inference system with trapezoidal MF 7 × 7. (b) Type 1 fuzzy inference system with triangular MF 7 × 7. (c) Type 1 fuzzy inference system with Gaussian MF 7 × 7. *(Contiued)*

The FIS editor [7, 26, 50, 65, 98, 104], the membership function editor, and the rule editor can all read and modify the FIS data, but the rule viewer and the surface viewer do not modify anything in the FIS data.

• One of the key issues in all fuzzy sets is how to determine fuzzy MFs.

• The membership function fully defines the fuzzy set.

• A membership function provides a measure of the degree of similarity of an element to a fuzzy set.

• Membership functions can take any form, but there are some common examples that appear in real applications.

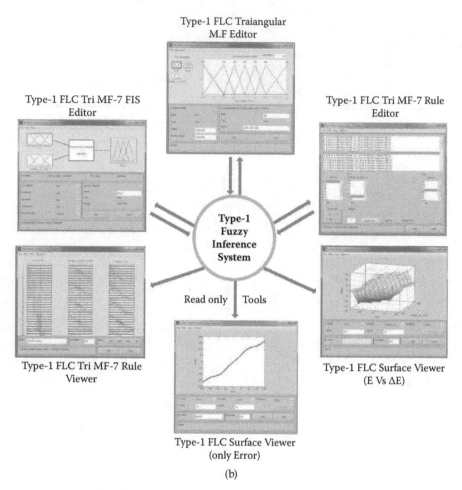

Type-1 FLC Traiangular M.F Editor

Type-1 FLC Tri MF-7 FIS Editor

Type-1 FLC Tri MF-7 Rule Editor

Type-1 Fuzzy Inference System

Read only | Tools

Type-1 FLC Tri MF-7 Rule Viewer

Type-1 FLC Surface Viewer (E Vs ΔE)

Type-1 FLC Surface Viewer (only Error)

(b)

FIGURE 3.11 (Continued)
(a) Type 1 fuzzy inference system with trapezoidal MF 7 × 7. (b) Type 1 fuzzy inference system with triangular MF 7 × 7. (c) Type 1 fuzzy inference system with Gaussian MF 7 × 7. *(Continued)*

- Membership functions can be either chosen by the user arbitrarily, based on the user's experience (membership function chosen by two users could be different, depending upon their experiences, perspectives, etc.), or designed using machine learning methods (e.g., artificial neural networks, genetic algorithms).

There are different shapes of membership functions [7, 26, 50, 65, 98, 104]: trapezoidal, triangular, sigmoidal, Gaussian, bell shaped, and so forth. Out of all these approaches, we have considered trapezoidal, triangular, and Gaussian MFs.

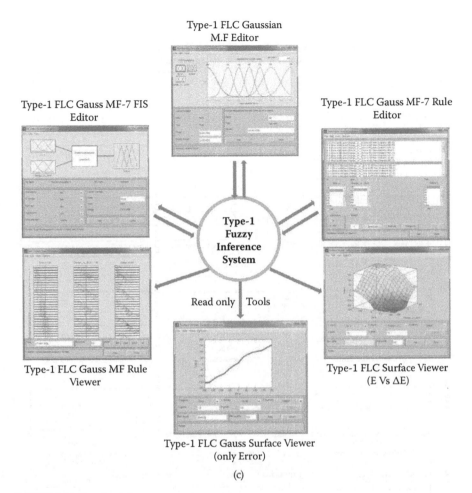

Type-1 FLC Gaussian M.F Editor

Type-1 FLC Gauss MF-7 FIS Editor

Type-1 FLC Gauss MF-7 Rule Editor

Type-1 Fuzzy Inference System

Read only | Tools

Type-1 FLC Gauss MF Rule Viewer

Type-1 FLC Surface Viewer (E Vs ΔE)

Type-1 FLC Gauss Surface Viewer (only Error)

(c)

FIGURE 3.11 (Continued)
(a) Type 1 fuzzy inference system with trapezoidal MF 7 × 7. (b) Type 1 fuzzy inference system with triangular MF 7 × 7. (c) Type 1 fuzzy inference system with Gaussian MF 7 × 7.

3.2 System Performance of Type 1 FLC-Based p-q Control Strategy with Different Fuzzy MFs Using MATLAB

Figures 3.12 through 3.14 present the details of source voltage, load current, compensation current, source current with filter, DC link voltage, and THD of the type 1 FLC-based p-q control strategy with different fuzzy MFs (trapezoidal, triangular, and Gaussian) using MATLAB/Simulink under balanced, unbalanced, and nonsinusoidal supply voltage conditions.

80 Power Quality Issues

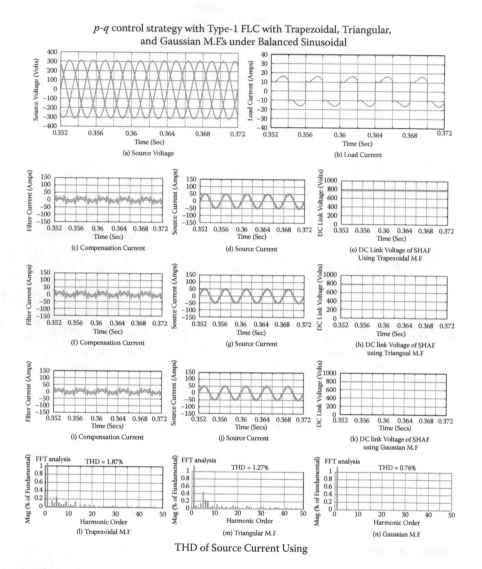

THD of Source Current Using

FIGURE 3.12

SHAF response using the *p-q* control strategy with type 1 FLC (trapezoidal, triangular, and Gaussian MF) under the balanced sinusoidal condition using MATLAB. (a) Source voltage, (b) load current, (c) compensation current using trapezoidal MF, (d) source current with filter using trapezoidal MF, (e) DC link voltage using trapezoidal MF, (f) compensation current using triangular MF, (g) source current with filter using triangular MF, (h) DC link voltage using triangular MF, (i) compensation current using Gaussian MF, (j) source current with filter using Gaussian MF, (k) DC link voltage using Gaussian MF, (l) THD of source current with trapezoidal MF, (m) THD of source current with triangular MF, and (n) THD of source current with Gaussian MF.

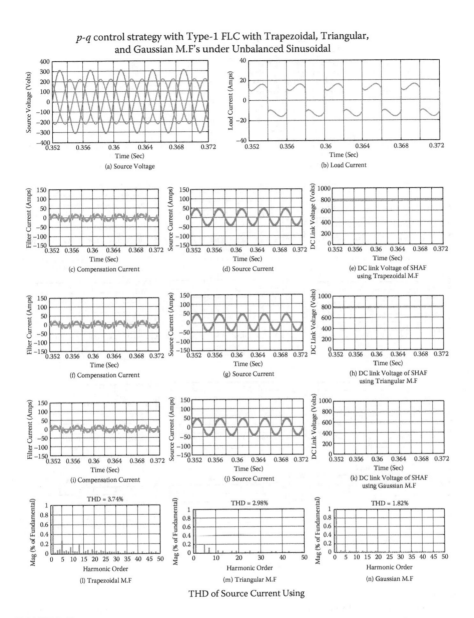

FIGURE 3.13

SHAF response using the *p-q* control strategy with type 1 FLC (trapezoidal, triangular, and Gaussian MF) under the unbalanced sinusoidal condition using MATLAB. (a) Source voltage, (b) load current, (c) compensation current using trapezoidal MF, (d) source current with filter using trapezoidal MF, (e) DC link voltage using trapezoidal MF, (f) compensation current using triangular MF, (g) source current with filter using triangular MF, (h) DC link voltage using triangular MF, (i) compensation current using Gaussian MF, (j) source current with filter using Gaussian MF, (k) DC link voltage using Gaussian MF, (l) THD of source current with trapezoidal MF, (m) THD of source current with triangular MF, and (n) THD of source current with Gaussian MF.

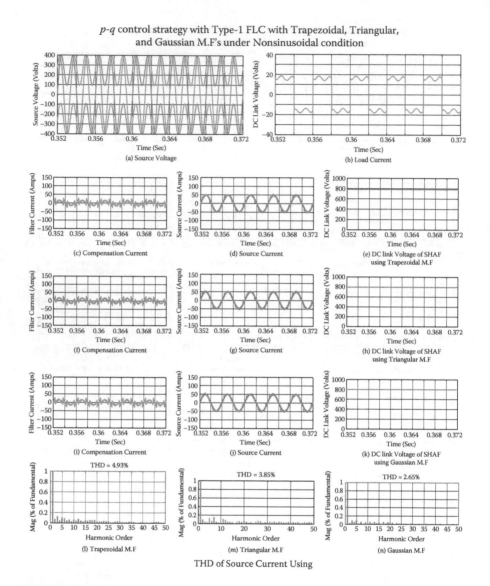

THD of Source Current Using

FIGURE 3.14

SHAF response using the *p-q* control strategy with type 1 FLC (trapezoidal, triangular, and Gaussian MF) under the nonsinusoidal condition using MATLAB. (a) Source voltage, (b) load current, (c) compensation current using trapezoidal MF, (d) source current with filter using trapezoidal MF, (e) DC link voltage using trapezoidal MF, (f) compensation current using triangular MF, (g) source current with filter using triangular MF, (h) DC link voltage using triangular MF, (i) compensation current using Gaussian MF, (j) source current with filter using Gaussian MF, (k) DC link voltage using Gaussian MF, (l) THD of source current with trapezoidal MF, (m) THD of source current with triangular MF, and (n) THD of source current with Gaussian MF.

Initially, the system performance is analyzed under balanced sinusoidal conditions, during which the type 1 FLC with all MFs is good enough at suppressing the harmonics, and the THDs are 1.87%, 1.27%, and 0.76%, respectively, using MATLAB. However, under unbalanced and nonsinusoidal conditions, the type 1 FLC with Gaussian MF shows superior performance over the type 1 FLC with the other two MFs.

The THD of the *p-q* control strategy using type 1 FLC with trapezoidal MF under the unbalanced condition is 3.74%, and under the nonsinusoidal condition it is 4.93%. The THD of the *p-q* control strategy using type 1 FLC with triangular MF under the unbalanced condition is about 2.98%, and under the nonsinusoidal condition it is about 3.85%.The THD of the *p-q* control strategy using type 1 FLC with Gaussian MF under the unbalanced condition is 1.82%, and under the nonsinusoidal condition it is 2.65%.

When the supply voltages are balanced and sinusoidal, the *p-q* control strategy using type 1 FLC with all membership functions (trapezoidal, triangular, and Gaussian) is converging to similar compensation characteristics. However, under unbalanced and nonsinusoidal conditions, the *p-q* control strategy using type 1 FLC with Gaussian MF shows superior performance over the type 1 FLC with the other two MFs.

The *p-q* control strategy using type 1 FLC with different MFs is unable to mitigate the harmonics perfectly; notches are observed in the source current. The main reason behind the notches is that the controller failed to track the current correctly, and thereby APF fails to compensate completely. So to mitigate the harmonics perfectly, one has to choose the perfect control strategy. So to avoid the difficulties occurring with the *p-q* control strategy, we have considered the I_d-I_q control strategy.

3.3 System Performance of Type 1 FLC-Based I_d-I_q Control Strategy with Different Fuzzy MFs Using MATLAB

Figures 3.15 through 3.17 present the details of source voltage, load current, compensation current, source current with filter, DC link voltage, and THD of the type 1 FLC-based I_d-I_q control strategy with different fuzzy MFs under balanced, unbalanced, and nonsinusoidal supply voltage conditions.

Initially, the system performance is analyzed under balanced sinusoidal conditions, during which the type 1 FLC with all MFs is good enough at suppressing the harmonics. The corresponding THDs are 1.15%, 0.97%, and 0.64%, respectively, using MATLAB.

The THD of the I_d-I_q control strategy using type 1 FLC with trapezoidal MF under the unbalanced condition is 2.26%, and under the nonsinusoidal

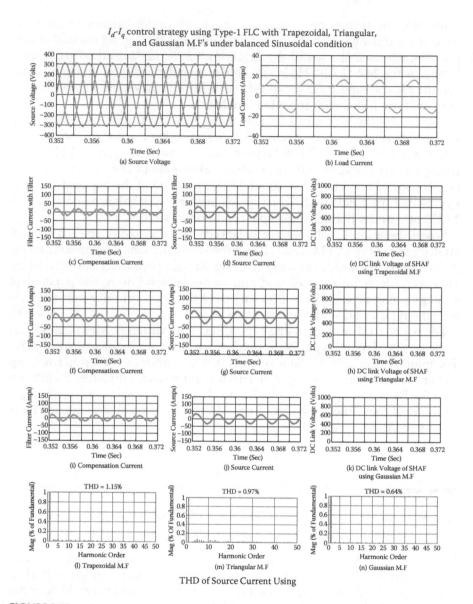

I_d-I_q control strategy using Type-1 FLC with Trapezoidal, Triangular, and Gaussian M.F's under balanced Sinusoidal condition

(a) Source Voltage

(b) Load Current

(c) Compensation Current

(d) Source Current

(e) DC link Voltage of SHAF using Trapezoidal M.F

(f) Compensation Current

(g) Source Current

(h) DC link Voltage of SHAF using Triangular M.F

(i) Compensation Current

(j) Source Current

(k) DC link Voltage of SHAF using Gaussian M.F

(l) Trapezoidal M.F

(m) Triangular M.F

(n) Gaussian M.F

THD of Source Current Using

FIGURE 3.15

SHAF response using the I_d-I_q control strategy with type 1 FLC (trapezoidal, triangular, and Gaussian MF) under the balanced sinusoidal condition using MATLAB. (a) Source voltage, (b) load current, (c) compensation current using trapezoidal MF, (d) source current with filter using trapezoidal MF, (e) DC link voltage using trapezoidal MF, (f) compensation current using triangular MF, (g) source current with filter using triangular MF, (h) DC link voltage using triangular MF, (i) compensation current using Gaussian MF, (j) source current with filter using Gaussian MF, (k) DC link voltage using Gaussian MF, (l) THD of source current with trapezoidal MF, (m) THD of source current with triangular MF, and (n) THD of source current with Gaussian MF.

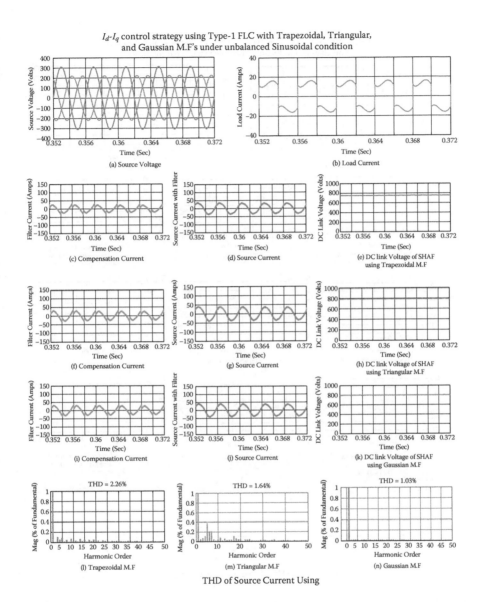

FIGURE 3.16

SHAF response using the I_d-I_q control strategy with type 1 FLC (trapezoidal, triangular, and Gaussian MF) under the unbalanced sinusoidal condition using MATLAB. (a) Source voltage, (b) load current, (c) compensation current using trapezoidal MF, (d) source current with filter using trapezoidal MF, (e) DC link voltage using trapezoidal MF, (f) compensation current using triangular MF, (g) source current with filter using triangular MF, (h) DC link voltage using triangular MF, (i) compensation current using Gaussian MF, (j) source current with filter using Gaussian MF, (k) DC link voltage using Gaussian MF, (l) THD of source current with trapezoidal MF, (m) THD of source current with triangular MF, and (n) THD of source current with Gaussian MF.

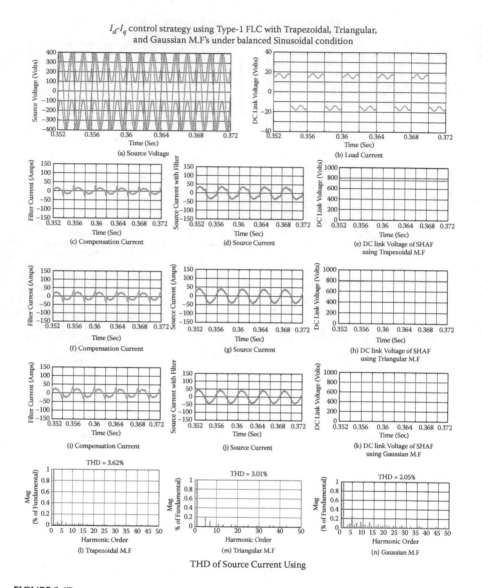

FIGURE 3.17

SHAF response using the I_d-I_q control strategy with type 1 FLC (trapezoidal, triangular, and Gaussian MF) under the nonsinusoidal condition using MATLAB. (a) Source voltage, (b) load current, (c) compensation current using trapezoidal MF, (d) source current with filter using trapezoidal MF, (e) DC link voltage using trapezoidal MF, (f) compensation current using triangular MF, (g) source current with filter using triangular MF, (h) DC link voltage using triangular MF, (i) compensation current using Gaussian MF, (j) source current with filter using Gaussian MF, (k) DC link voltage using Gaussian MF, (l) THD of source current with trapezoidal MF, (m) THD of source current with triangular MF, and (n) THD of source current with Gaussian MF.

condition it is 3.62%. The THD of the I_d-I_q control strategy using type 1 FLC with triangular MF under the unbalanced condition is 1.64%, and under the nonsinusoidal condition it is 3.01%. The THD of the I_d-I_q control strategy using type 1 FLC with Gaussian MF under the unbalanced condition is 1.03%, and under the nonsinusoidal condition it is 2.05%, using MATLAB.

3.4 System Performance of Type 1 FLC-Based p-q Control Strategy with Different Fuzzy MFs Using a Real-Time Digital Simulator

Figures 3.18 through 3.20 highlight the performance of the type 1 FLC-based p-q control strategy with different fuzzy MFs under balanced, unbalanced, and nonsinusoidal conditions using a real-time digital simulator.

Initially, the system performance is analyzed under balanced sinusoidal conditions, during which the type 1 FLC with all MFs are good enough at suppressing the harmonics, and the THDs of the p-q control strategy using a real-time digital simulator are 2.12%, 1.45%, and 1.23%. However, under unbalanced and nonsinusoidal conditions, the type 1 FLC with Gaussian MF shows superior performance over the type 1 FLC with the other two MFs.

The THD of the type 1 FLC with trapezoidal MF under the unbalanced condition is 3.98%, and under the nonsinusoidal condition it is 5.23%. The THD of the type 1 FLC with triangular MF under the unbalanced condition is about 3.27%, and under the nonsinusoidal condition it is 4.15%. The THD of the type 1 FLC with Gaussian MF under the unbalanced condition is 2.26%, and under the nonsinusoidal condition it is 2.89% using the p-q strategy with a real-time digital simulator.

3.5 System Performance of Type 1 FLC-Based I_d-I_q Control Strategy with Different Fuzzy MFs Using a Real-Time Digital Simulator

Figures 3.21 through 3.23 highlight the performance of the type 1 FLC-based I_d-I_q control strategy with different fuzzy MFs under balanced, unbalanced, and nonsinusoidal conditions using a real-time digital simulator.

Initially, the system performance is analyzed under balanced sinusoidal conditions, during which the type 1 FLC with all MFs is good enough at suppressing the harmonics. The respective THDs of the type 1 FLC-based I_d-I_q control strategy in a real-time digital simulator are 1.46%, 1.26%, and 0.93%.

p-q control strategy with Type-1 FLC with Trapezoidal, Triangular, and Gaussian
M.F's under balanced Sinusoidal

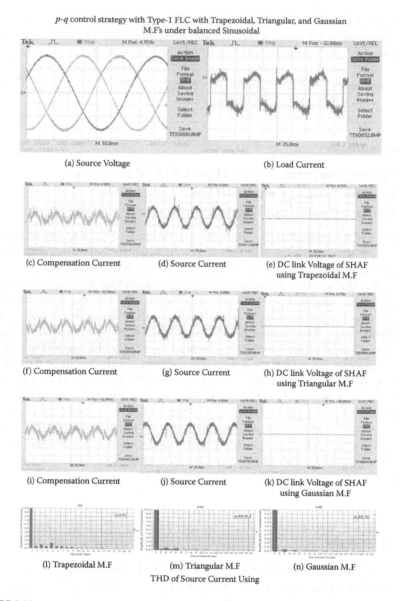

(a) Source Voltage   (b) Load Current

(c) Compensation Current   (d) Source Current   (e) DC link Voltage of SHAF
                                                      using Trapezoidal M.F

(f) Compensation Current   (g) Source Current   (h) DC link Voltage of SHAF
                                                      using Triangular M.F

(i) Compensation Current   (j) Source Current   (k) DC link Voltage of SHAF
                                                      using Gaussian M.F

(l) Trapezoidal M.F   (m) Triangular M.F   (n) Gaussian M.F
                     THD of Source Current Using

FIGURE 3.18
SHAF response using the *p-q* control strategy with type 1 FLC (trapezoidal, triangular, and Gaussian MF) under the balanced sinusoidal condition using a real-time digital simulator. (a) Source voltage, (b) load current (scale 20 A/div), (c) compensation current (scale 20 A/div) using trapezoidal MF, (d) source current (scale 30 A/div) with filter using trapezoidal MF, (e) DC link voltage using trapezoidal MF, (f) compensation current using triangular MF, (g) source current with filter using triangular MF, (h) DC link voltage using triangular MF, (i) compensation current using Gaussian MF, (j) source current with filter using Gaussian MF, (k) DC link voltage using Gaussian MF, (l) THD of source current with trapezoidal MF, (m) THD of source current with triangular MF, and (n) THD of source current with Gaussian MF.

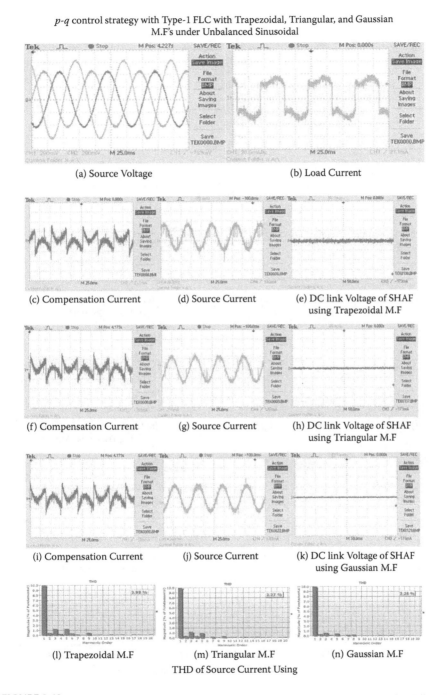

p-q control strategy with Type-1 FLC with Trapezoidal, Triangular, and Gaussian
M.F's under Unbalanced Sinusoidal

(a) Source Voltage (b) Load Current

(c) Compensation Current (d) Source Current (e) DC link Voltage of SHAF using Trapezoidal M.F

(f) Compensation Current (g) Source Current (h) DC link Voltage of SHAF using Triangular M.F

(i) Compensation Current (j) Source Current (k) DC link Voltage of SHAF using Gaussian M.F

(l) Trapezoidal M.F (m) Triangular M.F (n) Gaussian M.F
THD of Source Current Using

FIGURE 3.19
SHAF response using the p-q control strategy with type 1 FLC under the unbalanced sinusoidal condition using a real-time digital simulator.

p-q control strategy with Type-1 FLC with Trapezoidal, Triangular, and Gaussian
M.F's under Nonsinusoidal

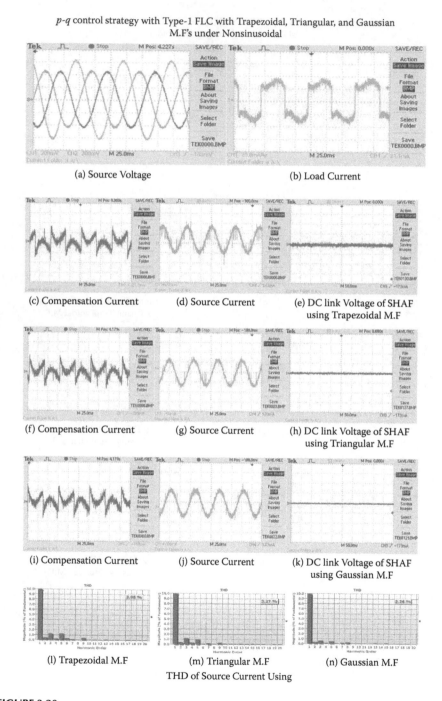

(a) Source Voltage

(b) Load Current

(c) Compensation Current

(d) Source Current

(e) DC link Voltage of SHAF
using Trapezoidal M.F

(f) Compensation Current

(g) Source Current

(h) DC link Voltage of SHAF
using Triangular M.F

(i) Compensation Current

(j) Source Current

(k) DC link Voltage of SHAF
using Gaussian M.F

(l) Trapezoidal M.F

(m) Triangular M.F

(n) Gaussian M.F

THD of Source Current Using

FIGURE 3.20
SHAF response using the *p-q* control strategy with type 1 FLC (trapezoidal, triangular, and
Gaussian MF) under the nonsinusoidal condition using a real-time digital simulator.

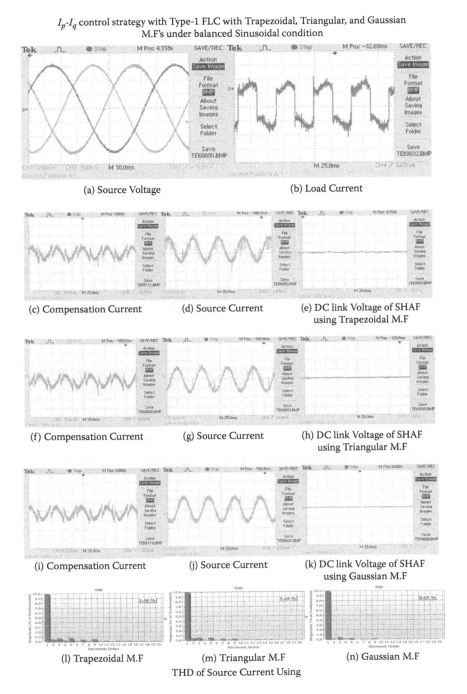

I_p-I_q control strategy with Type-1 FLC with Trapezoidal, Triangular, and Gaussian M.F's under balanced Sinusoidal condition

(a) Source Voltage

(b) Load Current

(c) Compensation Current

(d) Source Current

(e) DC link Voltage of SHAF using Trapezoidal M.F

(f) Compensation Current

(g) Source Current

(h) DC link Voltage of SHAF using Triangular M.F

(i) Compensation Current

(j) Source Current

(k) DC link Voltage of SHAF using Gaussian M.F

(l) Trapezoidal M.F

(m) Triangular M.F

(n) Gaussian M.F

THD of Source Current Using

FIGURE 3.21
SHAF response using the I_d-I_q control strategy with type 1 FLC under the balanced sinusoidal condition using a real-time digital simulator.

I_d-I_q control strategy with Type-1 FLC with Trapezoidal, Triangular, and Gaussian
M.F's under Unbalanced Sinusoidal

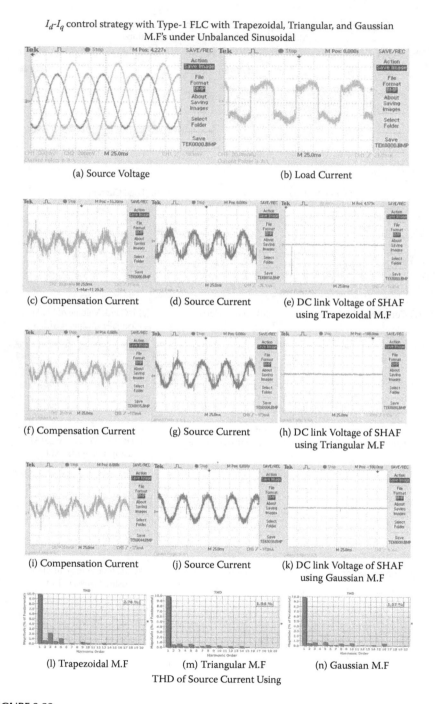

(a) Source Voltage   (b) Load Current

(c) Compensation Current   (d) Source Current   (e) DC link Voltage of SHAF
                                                          using Trapezoidal M.F

(f) Compensation Current   (g) Source Current   (h) DC link Voltage of SHAF
                                                          using Triangular M.F

(i) Compensation Current   (j) Source Current   (k) DC link Voltage of SHAF
                                                          using Gaussian M.F

(l) Trapezoidal M.F   (m) Triangular M.F   (n) Gaussian M.F
THD of Source Current Using

FIGURE 3.22
SHAF response using the I_d-I_q control strategy with type 1 FLC (trapezoidal, triangular, and
Gaussian MF) under the unbalanced sinusoidal condition using a real-time digital simulator.

I_d-I_q control strategy with Type-1 FLC with Trapezoidal, Triangular, and Gaussian
M.F's under Nonsinusoidal Condition

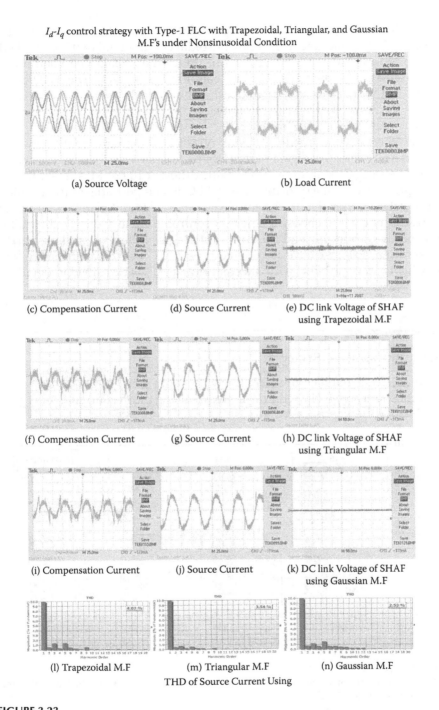

(a) Source Voltage

(b) Load Current

(c) Compensation Current

(d) Source Current

(e) DC link Voltage of SHAF
using Trapezoidal M.F

(f) Compensation Current

(g) Source Current

(h) DC link Voltage of SHAF
using Triangular M.F

(i) Compensation Current

(j) Source Current

(k) DC link Voltage of SHAF
using Gaussian M.F

(l) Trapezoidal M.F

(m) Triangular M.F

(n) Gaussian M.F

THD of Source Current Using

FIGURE 3.23
SHAF response using the I_d-I_q control strategy with type 1 FLC (trapezoidal, triangular, and
Gaussian MF) under the nonsinusoidal condition using a real-time digital simulator.

The THD of the type 1 FLC with trapezoidal MF under the unbalanced condition is 2.78%, and under the nonsinusoidal condition it is 4.07%. The THD of the type 1 FLC with triangular MF under the unbalanced condition is 1.94%, and under the nonsinusoidal condition it is 3.54%. The THD of the type 1 FLC with Gaussian MF under the unbalanced condition is 1.57%, and under the nonsinusoidal condition it is 2.53%, with the I_d-I_q control strategy using a real-time digital simulator.

With the I_d-I_q control strategy using type 1 FLC with different fuzzy MFs, the SHAF is able to mitigate harmonics in a better way than that of the p-q control strategy using type 1 FLC with different fuzzy MFs. Even though the I_d-I_q control strategy using type 1 FLC is able to mitigate the harmonics, notches are present in the source current. So to mitigate the harmonics perfectly, one has to choose the perfect controller. So to avoid the difficulties that occur with the p-q and I_d-I_q control strategies using type 1 FLC with different fuzzy MFs, we have considered type 2 FLC with different fuzzy MFs. In Chapter 4, type 2 FLC with different fuzzy MFs is explained in detail.

3.6 Comparative Study

Figures 3.24 through Figure 3.27 clearly illustrate the THD of the source current for shunt active filter control strategies using the PI controller and type 1 FLC with different fuzzy MFs.

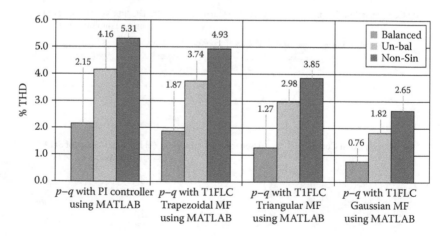

FIGURE 3.24
THD of source current for the p-q method using the PI controller and type 1 FLC with different fuzzy MFs using MATLAB.

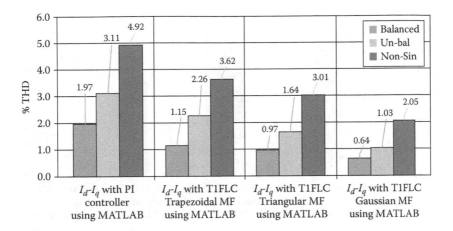

FIGURE 3.25
THD of the source current for the I_d-I_q method using the PI controller and type 1 FLC with different fuzzy MFs using MATLAB.

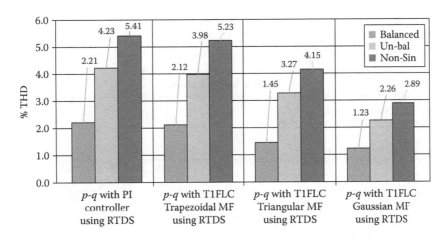

FIGURE 3.26
THD of source current for p-q method using the PI controller and type 1 FLC with different fuzzy MFs using a real-time digital simulator.

Figures 3.28 through Figure 3.31 clearly illustrate the amount of THD of the source current reduced from one controller to another controller for shunt active filter control strategies (p-q and I_d-I_q) under various source conditions (balanced, unbalanced, and nonsinusoidal) using MATLAB and a real-time digital simulator.

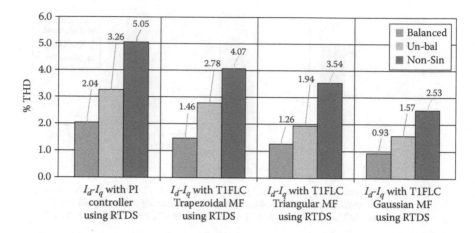

FIGURE 3.27
THD of source current for the I_d-I_q method using the PI controller and type 1 FLC with different fuzzy MFs using a real-time digital simulator.

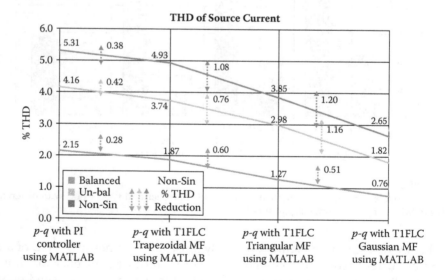

FIGURE 3.28
The amount of THD reduced for the PI controller and type 1 FLC with different fuzzy MFs using the *p-q* control strategy with MATLAB.

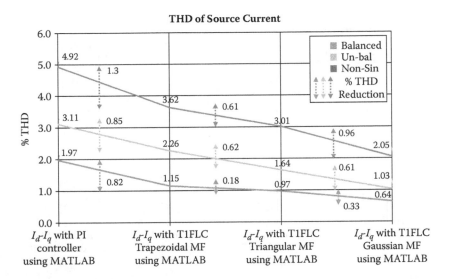

FIGURE 3.29
The amount of THD reduced for the PI controller and type 1 FLC with different fuzzy MFs using the I_d-I_q control strategy with MATLAB.

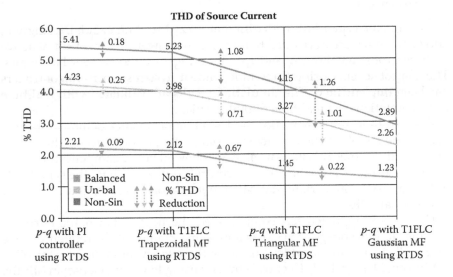

FIGURE 3.30
The amount of THD reduced for the PI controller and type 1 FLC with different fuzzy MFs using the *p-q* control strategy with a real-time digital simulator.

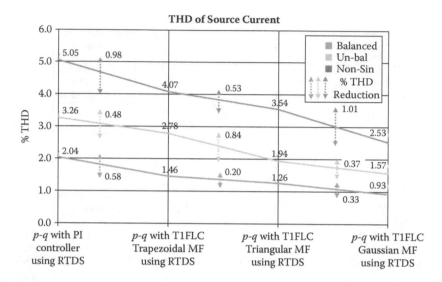

FIGURE 3.31
The amount of THD reduced for the PI controller and type 1 FLC with different fuzzy MFs using the I_d-I_q control strategy with a real-time digital simulator.

3.7 Summary

In this chapter, type 1 FLC with different fuzzy MFs (trapezoidal, triangular, and Gaussian) were developed to improve the power quality of shunt active filter control strategies (p-q and I_d-I_q) by mitigating the current harmonics. The control scheme using three independent hysteresis current controllers has been implemented. The simulation results are validated with real-time implementation on a real-time digital simulator.

The real-time implementation and simulation results demonstrate that even if the supply voltage is nonsinusoidal, the performance of the type 1 FLC-based I_d-I_q theory with Gaussian MF showed better compensation capabilities in terms of THD than the I_d-I_q control strategy with PI, the type 1 FLC with trapezoidal and triangular MF, the p-q theory with PI, and the type 1 FLC with all MFs.

The control approach has compensated the neutral harmonic currents, and the DC bus voltage of SHAF is almost maintained at the reference value under all disturbances, which confirms the effectiveness of the controller. While considering the I_d-I_q control strategy using FLC with Gaussian MF, the SHAF has been found to meet the IEEE 519–1992 standard recommendations on harmonic levels, making it easily adaptable to more severe constraints, such as highly distorted and unbalanced supply voltage.

With the I_d-I_q control strategy using type 1 FLC with different fuzzy MFs, the SHAF is able to mitigate current harmonics in a better way than that of the *p-q* control strategy using type 1 FLC with different fuzzy MFs. Even though the type 1 FLC-based I_d-I_q control strategy with different fuzzy MFs is able to mitigate the current harmonics, notches (small amount of harmonics) are present in the source current. So to mitigate the current harmonics perfectly, one has to choose the appropriate controller. Hence, to avoid the difficulties that occur with type 1 FLC-based *p-q* and I_d-I_q control strategies using different fuzzy MFs, we have considered type 2 FLC-based *p-q* and I_d-I_q control strategies with different fuzzy MFs.

4

Performance Analysis of SHAF Control Strategies Using Type 2 FLC with Different Fuzzy MFs

In Chapter 3, type 1 fuzzy logic controller (FLC)–based shunt active filter (SHAF) control strategies with different fuzzy membership functions (MFs) were discussed. Simulation and real-time results were also presented. Even though type 1 FLC-based shunt active filter control strategies with different fuzzy MFs are able to mitigate the harmonics, notches are present in the source current. So to mitigate the harmonics perfectly, one has to choose the perfect controller. In this chapter, the proposed type 2 FLC-based shunt active filter control strategies with different fuzzy MFs are introduced. Detailed simulation results using MATLAB®/Simulink® software are presented to support the feasibility of the proposed control strategies. With this approach, the compensation capabilities of SHAF are extremely good.

This chapter is organized as follows: Section 4.1 introduces the advantages of using type 2 FLC. Section 4.2 presents the detailed structure of type 2 FLC. Section 4.3 gives the details of the type 2 fuzzy inference system with different fuzzy MFs. Simulation results of the type 2 FLC-based p-q control strategy with different fuzzy MFs using MATLAB are presented in Section 4.4. Simulation results of the type 2 FLC-based I_d-I_q control strategy with different fuzzy MFs using MATLAB are presented in Section 4.5. Section 4.6 provides the comparative study, and finally, Section 4.7 gives concluding remarks.

4.1 Introduction to Type 2 FLC

The concept of fuzzy systems was introduced by Lotfi Zadeh in 1965 to process data and information affected by nonprobabilistic uncertainty/imprecision [82]. Soon after, it was proven to be an excellent choice for many applications, since it mimics human control logic. These were designed to represent mathematically the vagueness and uncertainty of linguistic problems [84], thereby obtaining formal tools to work with intrinsic imprecision in different types of problems. It is considered a generalization of the classic set theory [85]. Intelligent systems based on the fuzzy logic controller are fundamental tools for nonlinear complex system modeling [108]. The advantages of fuzzy logic controllers [7, 26, 50, 65, 98, 104] over conventional (proportional–integral (PI)) controllers are that they do not require an accurate mathematical model, can work with imprecise inputs, can handle nonlinearity, and are more robust than conventional PI controllers.

With the development of type 2 FLCs [105] and their ability to handle uncertainty [109], utilizing type 2 FLC has attracted much significance in recent years. It is an extension [110] of the concept of well-known type 1 fuzzy sets. A type 2 fuzzy set is characterized by a fuzzy membership function; that is, the membership grade for each element is also a fuzzy set in [0, 1], unlike a type 1 fuzzy set, where the membership grade is a crisp number in [0, 1]. The membership functions of type 2 fuzzy sets are three-dimensional and include a foot point of uncertainty (FOU) [107–116], which is the new third dimension of type 2 fuzzy sets. The FOU provides an additional degree of freedom to handle uncertainties.

Type 2 fuzzy sets [114] are used for modeling uncertainty and imprecision in a better way. The concept of type 2 fuzzy sets was first proposed by Lofti Zadeh in 1975 and is essentially fuzzy-fuzzy [115] sets where the fuzzy degree of membership is a type 1 fuzzy set. The new concepts were introduced by Mendel [117–122] and Liang [118, 122], allowing the characterization of a type 2 fuzzy set with a superior membership function and an inferior membership function. These two functions can each be represented by a type 1 fuzzy set membership function.

There are different sources of uncertainty [123] in the evaluation process. The five types of uncertainty that emerge from the imprecise knowledge natural state are

- *Measurement uncertainty*: It is the error on observed quantities.
- *Process uncertainty*: It is the dynamic randomness.
- *Model uncertainty*: It is the wrong specification of the model structure.
- *Estimate uncertainty*: It is the one that can appear from any of the previous uncertainties or a combination of them, and it is called inexactness and imprecision.
- *Implementation uncertainty*: It is the consequence of the variability that results from sorting politics, that is, incapacity to reach the exact strategic objective.

4.1.1 Why Type 2 FLCs?

Using type 2 fuzzy sets to represent the inputs/outputs of an FLC has many advantages when compared to the type 1 fuzzy sets. As the type 2 fuzzy sets membership functions are fuzzy and contain an FOU, they can model and handle the linguistic [125] and numerical uncertainties associated with the inputs and outputs of the FLC [125]. Using type 2 fuzzy sets to represent the FLC inputs and outputs will result in the reduction of the FLC rule base when compared to using type 1 fuzzy sets [126]. This is because the uncertainty represented in the FOU in type 2 fuzzy sets allows us to cover the same range as type 1 fuzzy sets with a smaller number of labels.

It has recently been shown that the extra degrees of freedom provided by the FOU enable a type 2 FLC [116–124] to produce outputs that cannot be achieved by type 1 FLCs [127] with the same number of membership functions [128]. A type 2 fuzzy set may give rise to an equivalent type 1 membership grade [129] that is negative or larger than unity. Each input and output will be represented by a large number of type 1 fuzzy sets, which are embedded in the type 2 fuzzy sets [130]. The use of such a large number of type 1 fuzzy sets to describe the input and output variables allows for a detailed description of the analytical control surface, as the addition of the extra levels of classification give a much smoother control surface and response. In addition, the type 2 FLC can be thought of as a collection of many different embedded type 1 FLCs.

4.2 The Structure of Type 2 FLC

From Figure 4.2, it can be seen that the structure of a type 2 FLC [131, 132] is very similar to the structure of a type 1 FLC (Figure 4.1), and the only difference exists in the output processing block. For a type 1 FLC [104, 132], the output processing block contains only a defuzzifier, but for a type 2 FLC, the output processing block includes a type reducer [104].

The fuzzy logic theory [133] is based on computation with fuzzy sets. While type 1 fuzzy sets allow for a fuzzy representation of a term to be made, the fact that the membership function of a type 1 set is crisp means that the

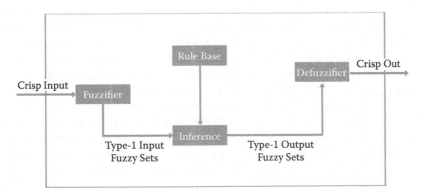

FIGURE 4.1
The architecture of type 1 FLC.

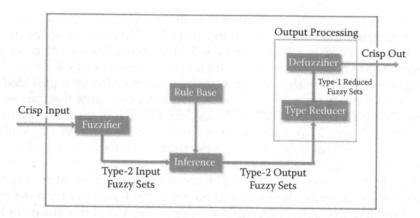

FIGURE 4.2
The architecture of type 2 FLC.

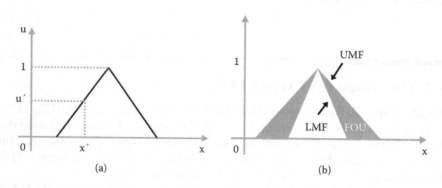

FIGURE 4.3
Membership functions of (a) type 1 FLC and (b) type 2 FLC.

degrees of the membership set are completely crisp—not fuzzy. In a type 1 fuzzy set (Figure 4.3a), the membership grade [134] for each element is a crisp number in [0, 1]. A type 2 fuzzy (Figure 4.3b) set is characterized by a three-dimensional membership function and a foot point of uncertainty (FOU).

In the design of the type 2 FLC [116, 135], the same configuration as that of the type 1 FLC is chosen. There are two inputs and a single output, and each input/output variable has the same seven linguistic variables, which are defined in Figure 4.3b. In this book, the Mamdani fuzzy inference system is used. Operation on the type 2 fuzzy set is identical with the operation on the type 1 fuzzy set. However, on the type 2 fuzzy system, fuzzy operation [136] is done at two type 1 membership functions [137], which limits the FOU, lower membership function (LMF), and upper membership function (UMF) to produce firing strengths. Hence, the membership value (or membership grade) for each element of this set is a fuzzy set in [0, 1].

A type 2 fuzzy set [138] is bounded from below by a lower membership function. The type 2 fuzzy set is bounded from above by an upper membership function. The area between the lower membership function and the upper membership is entitled the footprint of uncertainty (FOU) [139]. The new third dimension of type 2 fuzzy sets and the footprint of uncertainty provide additional degrees of freedom that make it possible to directly model and handle uncertainties. Hence, type 2 FLCs that use type 2 fuzzy sets in either their inputs or outputs have the potential to provide a suitable framework to handle the uncertainties in real-world environments. It is worth noting that a type 2 fuzzy set embeds a huge number of type 1 fuzzy sets [140]. A is a type 1 fuzzy set [141], and the membership grade [142] of $x \in X$ in A is $\mu_A(x)$, which is a crisp number in [0, 1]; a type 2 fuzzy set in X is $\tilde{A}$, and the membership grade of $x \in X$ in $\tilde{A}$ is $\mu_{\tilde{A}}(x)$, which is a type 1 fuzzy set in [0, 1].

A type 2 fuzzy set, denoted $\tilde{A}$, is characterized by a type 2 MF $\mu_{\tilde{A}}(x, u)$, where $x \in X$ and $u \in J_x \subseteq [0,1]$, that is,

$$\tilde{A} = \left\{ ((x, u), \mu_{\tilde{A}}(x, u)) \middle| \forall x \in X, \forall u \in J_x \subseteq [0, 1] \right\} \qquad (4.1)$$

in which

$$0 \le \mu_{\tilde{A}}(x, u) \le 1$$

where $\tilde{A}$ can also be expressed as

$$\tilde{A} = \int_{x \in X} \int_{u \in J_x} \mu_{\tilde{A}}(x, u)/(x, u) \qquad J_x \subseteq [0, 1] \qquad (4.2)$$

where $\iint$ denotes union over all admissible x and u. For a discrete universe of discourse, $\int$ is replaced by Σ.

$\tilde{A}$ can be reexpressed as

$$\tilde{A} = \left\{ (x, \mu_{\tilde{A}}(x)) \middle| \forall x \in X \right\} \qquad (4.3)$$

$$\tilde{A} = \int_{x \in X} \mu_{\tilde{A}}(x)/(x) = \int_{x \in X} \left[\int_{u \in J_x} f_x(u)/u \right] \middle/ x \qquad J_x \subseteq [0, 1] \qquad (4.4)$$

If X and J_x are both discrete, then

$$\tilde{A} = \sum_{x \in X} \left[\sum_{u \in J_x} f_x(u)/u \right] \bigg/ x$$

$$\tilde{A} = \sum_{i=1}^{N} \left[\sum_{u \in J_x} f_{x_i}(u)/u \right] \bigg/ x_i \qquad (4.5)$$

$$\tilde{A} = \sum_{K=1}^{M_1} \left[\sum_{u \in J_{x_1}} f_{x_1}(u_{1k})/u_{1k} \right] \bigg/ x_1 + \sum_{K=2}^{M_2} \left[\sum_{u \in J_{x_2}} f_{x_2}(u_{2k})/u_{2k} \right] \bigg/ x_2 + \dots$$

$$\sum_{K=1}^{M_N} \left[\sum_{u \in J_{x_N}} f_{x_N}(u_{Nk})/u_{Nk} \right] \bigg/ x_N \qquad (4.6)$$

The discretization [29] along each u_{ik} does not have to be the same, which is why we have shown a different upper sum for each of the bracketed terms. If, however, the discretization along each u_{ik} is the same, then $M_1 = M_2 = \dots = M_N = M$. Uncertainty in the primary memberships of a type 2 fuzzy set $\tilde{A}$ consists of a bounded region that we call the footprint of uncertainty [143]. It is the union of all primary memberships, that is,

$$FOU(\tilde{A}) = \bigcup_{x \in X} J_x \qquad (4.7)$$

4.3 Type 2 Fuzzy Inference System with Different Fuzzy MFs

Figure 4.4 shows the type 2 FLS [104, 132] with different MFs. It consists of

- Type 2 fuzzy inference system (type 2 FIS) editor
- Type 2 fuzzy membership function editor
- Type 2 fuzzy rule editor
- Type 2 fuzzy rule viewer
- Type 2 fuzzy surface viewer
- Type 2 fuzzy reduced surface viewer

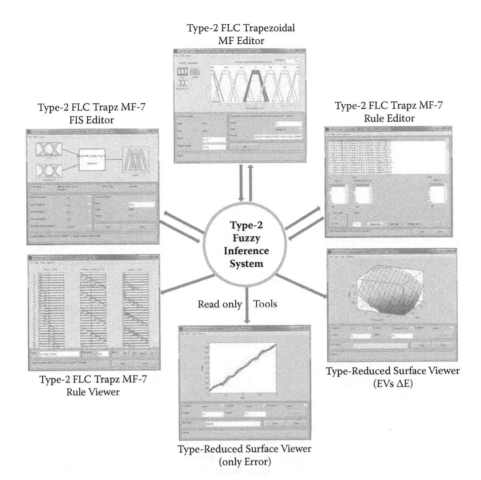

FIGURE 4.4

(a) Type 2 fuzzy inference system with trapezoidal MF 7 × 7. (b) Type 2 fuzzy inference system with triangular MF 7 × 7. (c) Type 2 fuzzy inference system with Gaussian MF 7 × 7. *(Continued)*

Figure 4.4a–c shows the type 2 FIS implementation with different fuzzy MFs using MATLAB. In this FIS we have designed [104, 132]

- Number of inputs and outputs (dual input and single output)
- Number of rules (49 rules)
- Type of membership function (trapezoidal, triangular, and Gaussian)
- Number of membership functions (seven)
- Type of implication (Mamdani max-min operation)
- Type of defuzzification method (centroid of area method)

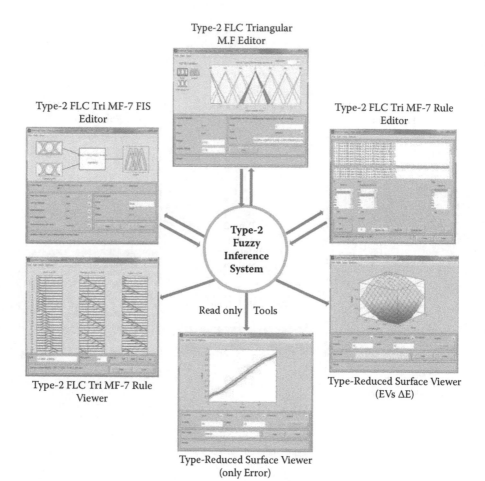

Type-2 FLC Triangular
M.F Editor

Type-2 FLC Tri MF-7 FIS
Editor

Type-2 FLC Tri MF-7 Rule
Editor

Type-2
Fuzzy
Inference
System

Read only | Tools

Type-2 FLC Tri MF-7 Rule
Viewer

Type-Reduced Surface Viewer
(EVs ΔE)

Type-Reduced Surface Viewer
(only Error)

FIGURE 4.4 (Continued)
(a) Type 2 fuzzy inference system with trapezoidal MF 7 × 7. (b) Type 2 fuzzy inference system with triangular MF 7 × 7. (c) Type 2 fuzzy inference system with Gaussian MF 7 × 7. *(Continued)*

The type 2 FIS [104] editor handles the high-level issues for the system. How many input and output variables? What are their names? The type 2 FIS doesn't limit the number of inputs. However, the number of inputs may be limited by the available memory of the PC (personal computer). If the number of inputs is too large, or the number of type 2 membership functions [38] is too big, then it may also be difficult to analyze the type 2 FIS using the other GUI tools. The type 2 membership function editor is used to define the shapes of all the type 2 membership functions associated with each variable. The type 2 rule editor is for editing the list of rules that define the behavior of the systems.

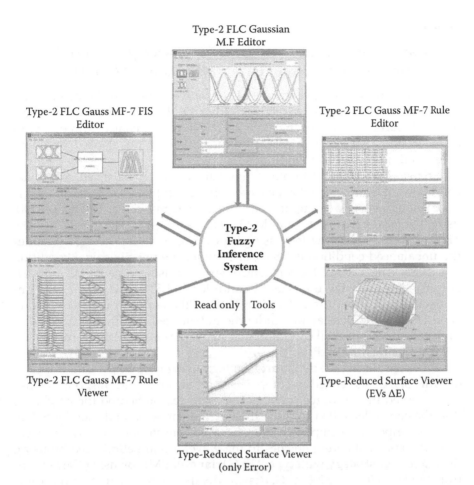

Type-2 FLC Gaussian
M.F Editor

Type-2 FLC Gauss MF-7 FIS
Editor

Type-2 FLC Gauss MF-7 Rule
Editor

Type-2
Fuzzy
Inference
System

Read only | Tools

Type-2 FLC Gauss MF-7 Rule
Viewer

Type-Reduced Surface Viewer
(EVs ΔE)

Type-Reduced Surface Viewer
(only Error)

FIGURE 4.4 (Continued)
(a) Type 2 fuzzy inference system with trapezoidal MF 7 × 7. (b) Type 2 fuzzy inference system with triangular MF 7 × 7. (c) Type 2 fuzzy inference system with Gaussian MF 7 × 7.

The type 2 rule viewer and the surface viewer or type-reduced surface viewer are used for looking at, as opposed to editing, the type 2 FIS [132]. They are strictly read-only tools. The type 2 rule viewer is a MATLAB-based display of the type 2 fuzzy inference diagram shown in Figure 4.4. Used as a diagnostic, it can show which rules are active, or how individual type 2 membership function shapes are influencing the results. The surface viewer is used to display the dependency of one of the outputs on any one or two of the inputs; that is, it generates and plots an output surface map for the system.

4.4 System Performance of Type 2 FLC-Based p-q Control Strategy with Different Fuzzy MFs Using MATLAB

Figures 4.5 through 4.7 give the details of source voltage, load current, compensation current, source current with filter, DC link voltage, and total harmonic distortion (THD) of the type 2 FLC-based p-q control strategy with different fuzzy MFs using MATLAB under balanced, unbalanced, and nonsinusoidal supply voltage conditions.

Initially, the system performance is analyzed under balanced sinusoidal conditions, during which it is observed that the type 2 FLC with all MFs is good enough at suppressing the harmonics. The respective THDs of SHAF using the p-q control strategy in MATLAB are 0.93%, 0.78%, and 0.45%. The THD of the p-q control strategy using type 2 FLC with trapezoidal MF under the unbalanced condition is 1.86%, and under the nonsinusoidal condition it is 2.95%. The THD of the p-q control strategy using type 2 FLC with triangular MF under the unbalanced condition is 1.53%, and under the nonsinusoidal condition it is 2.66%. The THD of the p-q control strategy using type 2 FLC with Gaussian MF under the unbalanced condition is 0.85%, and under the nonsinusoidal condition it is 1.29%. These simulations results are obtained using MATLAB.

While considering the p-q control strategy using type 2 FLC with trapezoidal MF, SHAF succeeded in compensating harmonic currents, but notches are observed in the source current. The main reason behind the notches is that the controller failed to track the current correctly, and thereby SHAF fails to compensate completely. It is observed that the source current waveform is somewhat better; notches in the waveform are eliminated by using the p-q control strategy with type 2 FLC triangular MF. By using the p-q control strategy with type 2 FLC Gaussian MF, the source current waveform is improved in quality and notches in the waveform are minimized.

Even though the p-q control strategy using type 2 FLC with different fuzzy MFs is able to mitigate the harmonics, notches are observed in the source current. So to mitigate the harmonics perfectly, one has to choose the perfect method. To avoid the difficulties that occur with the p-q control strategy, we have considered the type 2 FLC-based I_d-I_q control strategy.

4.5 System Performance of Type 2 FLC-Based I_d-I_q Control Strategy with Different Fuzzy MFs Using MATLAB

Figures 4.8 through 4.10 give the details of source voltage, load current, compensation current, source current with filter, DC link voltage, and total harmonic distortion (THD) of the type 2 FLC-based I_d-I_q control strategy with

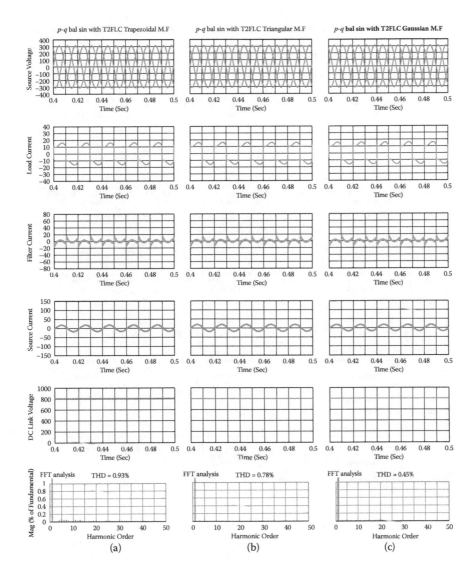

FIGURE 4.5
SHAF response using the *p-q* control strategy with type 2 FLC under the balanced sinusoi-dal condition using MATLAB. (a) Trapezoidal MF, (b) triangular MF, and (c) Gaussian MF. (i) Source voltage, (ii) load current, (iii) compensation current, (iv) source current with filter, (v) DC link voltage, and (vi) THD of the source current.

different fuzzy MFs using MATLAB under balanced, unbalanced, and non-sinusoidal supply voltage conditions.

Initially, the system performance is analyzed under balanced sinusoidal conditions, during which the type 2 FLC with all MFs are good enough at suppressing the harmonics. The respective THDs of SHAF in MATLAB are 0.48%, 0.35%, and 0.27%.

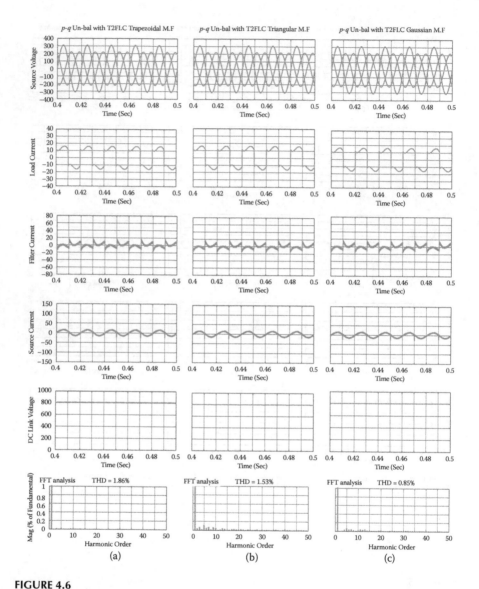

FIGURE 4.6
SHAF response using the *p-q* control strategy with type 2 FLC under the unbalanced sinusoidal condition using MATLAB.

The THD of the I_d-I_q control strategy using type 2 FLC with trapezoidal MF under the unbalanced condition is 1.23%, and under the nonsinusoidal condition it is 2.46%. The THD of the I_d-I_q control strategy using type 2 FLC with triangular MF under the unbalanced condition is 0.96%, and under the nonsinusoidal condition it is 1.83%. The THD of the I_d-I_q control strategy using type 2 FLC with Gaussian MF under the unbalanced condition is

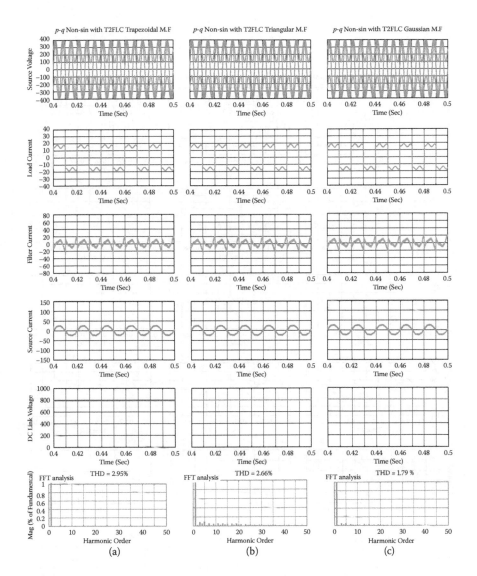

FIGURE 4.7
SHAF response using the *p-q* control strategy with type 2 FLC under the nonsinusoidal condition using MATLAB. (a) Trapezoidal MF, (b) triangular MF, and (c) Gaussian MF. (i) Source voltage, (ii) load current, (iii) compensation current, (iv) source current with filter, (v) DC link voltage, and (vi) THD of the source current.

0.66%, and under the nonsinusoidal condition it is 1.44%. The simulations are performed using MATLAB.

When the supply voltages are balanced and sinusoidal, the I_d-I_q control strategy using type 2 FLC, with all membership functions (trapezoidal, triangular, and Gaussian), is converging to the same compensation characteristics.

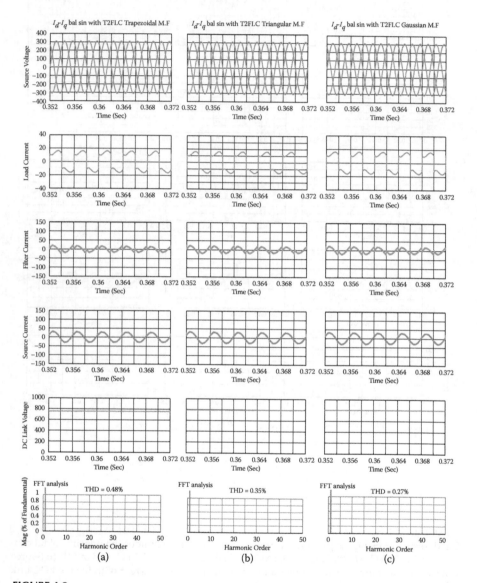

FIGURE 4.8
SHAF response using the I_d-I_q control strategy with type 2 FLC under the balanced sinusoidal condition using MATLAB. (a) Trapezoidal MF, (b) triangular MF, and (c) Gaussian MF. (i) Source voltage, (ii) load current, (iii) compensation current, (iv) source current with filter, (v) DC link voltage, and (vi) THD of the source current.

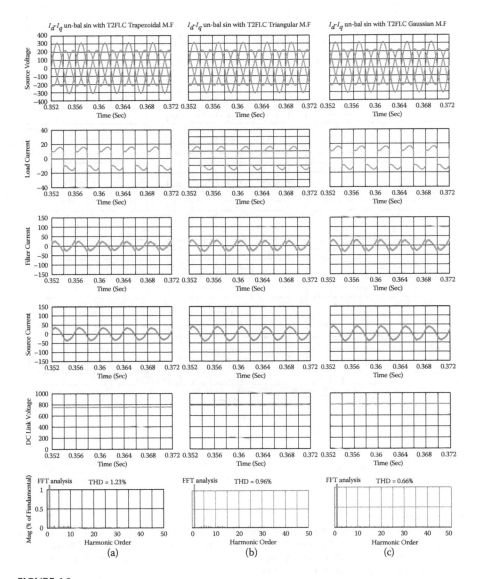

FIGURE 4.9
SHAF response using the I_d-I_q control strategy with type 2 FLC under the unbalanced sinusoidal condition using MATLAB.

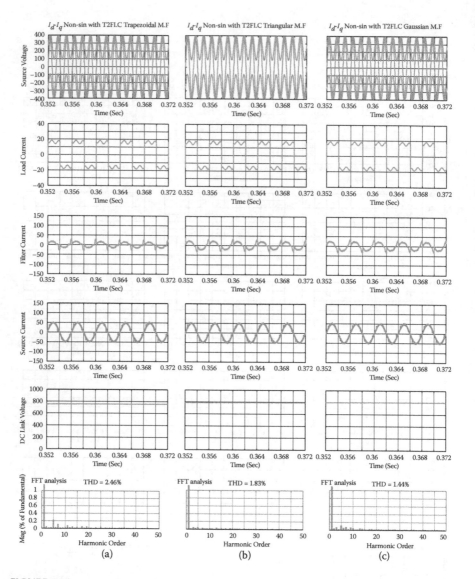

FIGURE 4.10
SHAF response using the I_d-I_q control strategy with type 2 FLC under the nonsinusoidal condition using MATLAB. (a) Trapezoidal MF, (b) triangular MF, and (c) Gaussian MF. (i) Source voltage, (ii) load current, (iii) compensation current, (iv) source current with filter, (v) DC link voltage, and (vi) THD of the source current.

However, under unbalanced and nonsinusoidal conditions, the I_d-I_q control strategy using type 2 FLC with Gaussian MF shows superior performance over the type 2 FLC with the other two MFs.

While considering the I_d-I_q control strategy using type 2 FLC with different fuzzy MFs, SHAF succeeds in compensating harmonic currents. It is observed that the quality of source current waveforms is extremely good, and notches in the waveform are eliminated.

4.6 Comparative Study

Figures 4.11 and 4.12 illustrate the THD of the source current for *p-q* and I_d-I_q control strategies using the PI controller, type 1 FLC, and type 2 FLC with different fuzzy MFs under various source voltage conditions (balanced, unbalanced, and nonsinusoidal), respectively.

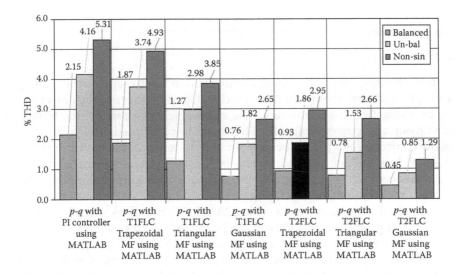

FIGURE 4.11
Bar graph indicating the THD of the source current for the *p-q* control strategy using the PI controller, type 1 FLC, and type 2 FLC with different fuzzy MFs using MATLAB.

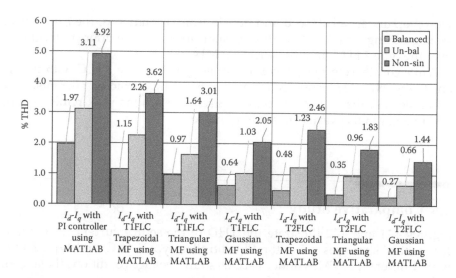

FIGURE 4.12
Bar graph indicating the THD of the source current for the I_d-I_q control strategy using the PI controller, type 1 FLC, and type 2 FLC with different fuzzy MFs using MATLAB.

4.7 Summary

In this chapter, type 2 FLC-based shunt active filter control strategies (p-q and I_d-I_q) with different fuzzy MFs are developed to improve the power quality by mitigating the harmonics. The control scheme using three independent hysteresis current controllers has been implemented. The performance of the control strategies has been evaluated, in terms of harmonic mitigation and DC link voltage regulation. The proposed SHAF with different fuzzy MFs (trapezoidal, triangular, and Gaussian) is able to eliminate the uncertainty in the system, and SHAF gains outstanding compensation abilities. The detailed simulation results using MATLAB/Simulink software are presented to support the feasibility of the proposed control strategies.

The simulation results showed that even if the supply voltage is nonsinusoidal, the performance of the type 2 FLC-based I_d-I_q control strategy with Gaussian MF showed better compensation capabilities in terms of THD than the I_d-I_q theory with PI, type 1 FLC with all MFs, and type 2 FLC with trapezoidal and triangular MFs, and also the p-q theory with PI, type 1 FLC, and type 2 FLC with all MFs. While considering the I_d-I_q control strategy using type 2 FLC with different fuzzy MFs, SHAF succeeded in compensating harmonic currents. It is observed that the quality of source current waveforms is extremely good, and notches in the waveform are also eliminated.

The control approach has compensated the neutral harmonic currents, and the DC bus voltage of SHAF is almost maintained at the reference value under all disturbances, which confirms the effectiveness of the controller. While considering the I_d-I_q control strategy using type 1 FLC and type 2 FLC and the p-q control strategy with type 2 FLC, the SHAF has been found to meet the IEEE 519–1992 standard recommendations on harmonic levels, making it easily adaptable to more severe constraints, such as highly distorted or unbalanced supply voltage.

5

Introduction to RT-LAB and Real-Time Implementation of Type 2 FLC-Based SHAF Control Strategies

In Chapter 4, the proposed type 2 fuzzy logic controller was discussed, and simulation results were also presented. It was concluded that the proposed type 2 FLC-based shunt active filter (SHAF) control strategies are suitable for mitigation of harmonics presented in the system even if the supply voltage is nonsinusoidal or distorted. So in this chapter, the proposed type 2 FLC-based shunt active filter control strategies are verified with a real-time digital simulator (OPAL-RT) to validate the proposed research.

This chapter is organized as follows: Section 5.1 introduces the role and advantages of using real-time simulation by answering three fundamental questions: What is real-time simulation? Why is it needed? How does it work? Section 5.2 provides the details of the evolution of real-time simulators. The details of the RT-LAB simulator architecture are given in Section 5.3, while Section 5.4 provides the details of how RT-LAB works, and Section 5.5 details the OP5142 configuration. Real-time results of type 2 FLC-based p-q and I_d-I_q control strategies with different fuzzy MFs using a real-time digital simulator (OPAL-RT) are presented in Sections 5.6 and 5.7. Section 5.8 provides the comparative study, and finally, Section 5.9 gives the concluding remarks.

5.1 Introduction to RT-LAB

Simulation tools have been widely used for the design and enhancement of electrical systems since the mid-twentieth century. The evolution of simulation tools has progressed in step with the evolution of computing technologies [132]. In recent years, computing technologies have upgraded dramatically in performance and become extensively available at a steadily decreasing cost. Consequently, simulation tools have also seen dramatic performance gains, and there is a steady decrease in cost. Researchers and engineers now have access to affordable, high-performance simulation tools that were previously too costly, except for the largest manufacturers and utilities. RT-LAB [7, 50, 65, 98, 104, 132, 144–157], fully integrated with MATLAB®/Simulink®, is the open real-time simulation software environment that has revolutionized the way model-based design is performed. RT-LAB's flexibility and scalability allow it to be used in virtually any simulation or control system application, and to add computing power to simulations where and when it is needed.

This simulator was developed with the aim of meeting the transient simulation needs of electromechanical drives and electric systems while solving the limitations of traditional real-time simulators. It is based on a central principle: the use of extensively available, user-friendly, highly competitive commercial products (PC platform, Simulink). The real-time simulator consists of two main tools: a real-time distributed simulation package (RT-LAB) [144] for the execution of Simulink block diagrams on a PC cluster, and algorithmic toolboxes designed for the fixed-time-step simulation of stiff electric circuits and their controllers. Real-time simulation [145] and hardware-in-the-loop (HIL) applications [146–148] are progressively recognized as essential tools for engineering design, especially in power electronics and electrical systems [149].

5.1.1 Why Use Real-Time Simulation?

Gain time

- Allows test engineers to gain time in the testing process
- Finds problems at an earlier stage in the design process
- Proceeds to a device design while the actual system is not physically available

Lower cost

- Reduces enormous cost on testing a new device under real conditions
- Tests many possible configurations without physical modification
- Reduces total cost over the entire project and system life cycle

Increase test functionalities

- Fakes and tests all possible scenarios that could happen in real life in a secure and simulated environment
- Has high flexibility by being able to modify all parameters and signals of the test system at a glance
- Has an automatic test script in order to run tests 24 hours a day, 7 days a week

5.1.2 What Is a Real-Time Simulation?

Fixed-step solvers solve the model at regular time intervals from the beginning to the end of the simulation. The size of the interval is known as the step size: *Ts*. Generally, decreasing *Ts* increases the accuracy of the results while increasing the time required to simulate the system [150].

In a real-time system [151], we define the *time step* as a predetermined amount of time (e.g., *Ts* = 10 μs, 1 ms, or 5 ms). Inside this amount of time, the processor has to read input signals, such as sensors, to perform all necessary calculations, such as control algorithms, and write all outputs, such as control actuators.

Inputs or outputs are the highest-frequency sampling consideration; generally, decreasing the time step increases the accuracy of the results, while increasing the time required to simulate the system. The rule of thumb is to have around 10 to 20 samples per period for an AC signal for a 1 kHz signal: $1/(20 * 1 \text{ kHz}) = 50$ μs.

5.2 Evolution of Real-Time Simulators

Simulator technology has evolved from physical/analog simulators (*high-voltage direct current* (HVDC) simulators and *transient network analyzers* (TNAs)) for electromagnetic transient (EMT) and protection and control studies to hybrid TNA/analog/digital simulators capable of studying EMT behavior to fully digital real-time simulators, as illustrated in Figure 5.1. Physical simulators served their purpose well. However, they were very

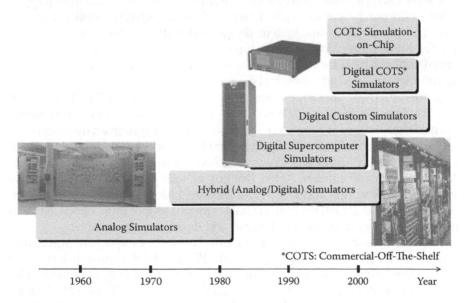

FIGURE 5.1
Evolution of RT-LAB simulator.

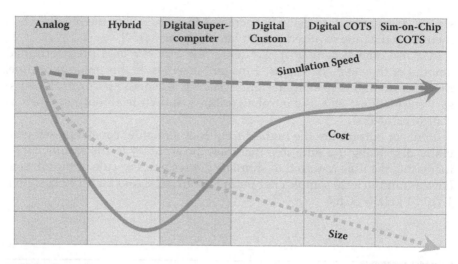

FIGURE 5.2
Speed, cost, and size of RT-LAB simulators.

large, expensive, and required highly skilled technical teams to handle the
tedious jobs of setting up networks and maintaining extensive inventories
of complex equipment [150]. With the development of microprocessor and
floating-point digital signal processor (DSP) technologies, physical simulators have been gradually replaced with fully digital real-time simulators.
Figure 5.2 gives the details of speed, cost, and size of RT-LAB simulators

Commercial off the shelf (COTS)–based high-end real-time simulators equipped
with multicore processors have been used in aerospace, robotics, automotive, and power electronic system design and testing for a number of years
[6]. Recent advancements in multicore processor technology have made such
simulators available for the simulation of EMT expected in large-scale power
grids, microgrids, wind farms, and power systems installed in all-electric ships
and aircraft. These simulators, operating under Windows, LINUX, and standard real-time operating systems (RTOS), have the potential to be compatible
with a large number of commercially available power system analysis software
tools, such as PSS/E (Power System Simulator for Engineering), EMTP-RV
(Electromagnetic Transients Program—Restructured Version), and PSCAD
(Power Systems Computer Aided Design), as well as multidomain software
tools such as Simulink and DYMOLA. The integration of multidomain simulation tools with electrical simulators enables the analysis of interactions between
electrical, power electronic, mechanical, and fluid dynamic systems.

The latest trend in real-time simulation consists of exporting simulation
models to a *field-programmable gate array* (FPGA). This approach has many
advantages. First, computation time within each time step is almost independent of system size because of the parallel nature of FPGAs [152]. Second,
overruns cannot occur once the model is running and timing constraints are
met. Last, but most important, the simulation time step can be very small, in

the order of 250 ns. There are still limitations on model size since the number of gates is limited in FPGAs. However, this technique holds promise.

5.3 RT-LAB Simulator Architecture

5.3.1 Block Diagram and Schematic Interface

The present real-time electric simulator is based on the RT-LAB real-time, distributed simulation platform; it is optimized to run Simulink in real time, with efficient fixed-step solvers, on a PC cluster [153]. Based on COTS nonproprietary PC components, RT-LAB is a flexible real-time simulation platform, for the automatic implementation of system-level, block diagram models on standard PCs. It uses the popular MATLAB/Simulink as a front end for editing and viewing graphic models in block diagram format. The block diagram models become the source from which code can be automatically generated, manipulated, and downloaded onto target processors (Pentium and Pentium compatible) for real-time or distributed simulation.

5.3.2 Inputs and Outputs

A requirement for real-time HIL applications is interfacing with real-world hardware devices, controller or physical plant alike. In the RT-LAB real-time simulator, input/output (I/O) interfaces are configured through custom blocks, supplied as a Simulink toolbox. The engineer merely needs to drag and drop the blocks to the graphic model and connect the inputs and outputs to these blocks, without worrying about low-level driver programming. RT-LAB manages the automatic generation of I/O drivers and models code to direct the model's data flow onto the physical I/O cards.

5.3.3 Simulator Configuration

In a typical configuration (Figure 5.3), the RT-LAB simulator consists of

- One or more target PCs (computation nodes). One of the PCs (master) manages the communication between the hosts and the targets and the communication between all other target PCs. The targets use the Red Hat real-time operating system.
- One or more host PCs allowing multiple users to access the targets. One of the hosts has full control of the simulator, while other hosts, in read-only mode, can receive and display signals from the real-time simulator.
- I/Os of various types (analog in and out, digital in and out, pulse width modulation (PWM) in and out, timers, encoders, etc.).

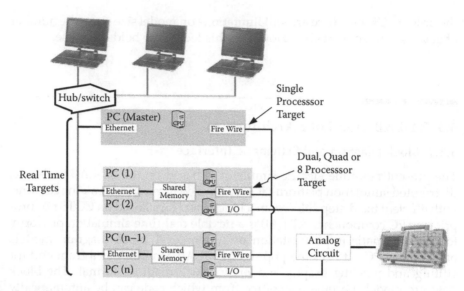

FIGURE 5.3
RT-LAB simulator architecture.

The simulator uses the following communication links:

- Ethernet connection (100 Mb/s) between the hosts and target PCs
- Ethernet connection between target nodes, allowing parallel computation of models with low and medium step size (in the millisecond range), or for free running, on a real-time simulation
- Fast shared-memory communication between processors on the same motherboard (dual, quad, or eight processors)
- Fast IEEE 1394 (FireWire) communication links (400 Mb/s) between target PCs for parallel simulation of models with small step sizes (down to 20 μs) and tight communication constraints (power systems, electric drive control, etc.)

5.4 How RT-LAB Works

RT-LAB allows the user to readily convert Simulink models, via Real-Time Workshop (RTW), and then to conduct real-time simulation of those models executed on multiple target computers equipped with multicore PC processors [147]. This is used particularly for HIL and rapid control prototyping applications [155, 156]. RT-LAB transparently handles synchronization, user interaction, real-world interfacing using I/O boards, and data exchanges for seamless distributed execution.

5.4.1 Single-Target Configuration

In this configuration (Figures 5.4 and 5.5), typically used for rapid control prototyping, a single computer runs the plant simulation or control logic. One or more hosts may connect to the target via an Ethernet link. The target uses QNX or Linux as the RTOS [157] for fast simulation or for applications where real-time performance is required. RT-LAB [7, 50, 65, 98, 104, 132] used Red Hat ORT, which is the standard Red Hat distribution package with an optimized set of parameters to reach real-time performance, enabling a model time step as low as 10 μs on multicore processors to be reached.

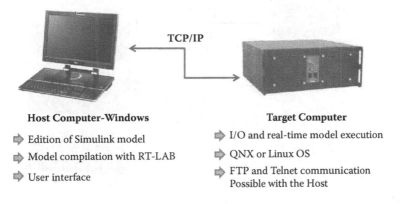

Host Computer-Windows

⇨ Edition of Simulink model
⇨ Model compilation with RT-LAB
⇨ User interface

Target Computer

⇨ I/O and real-time model execution
⇨ QNX or Linux OS
⇨ FTP and Telnet communication Possible with the Host

FIGURE 5.4
RT-LAB simulator with a single-target system.

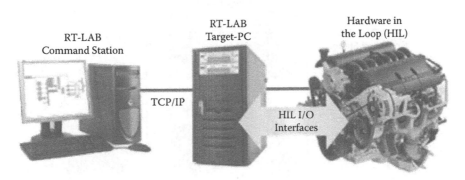

FIGURE 5.5
RT-LAB simulator with a single-target system and HIL.

5.4.2 Distributed Target Configuration

The distributed configuration (Figures 5.6 and 5.7) allows for complex models to be distributed over a cluster of multicore PCs [147] running in parallel.

The target nodes in the cluster communicate between each other with low-latency protocols such as FireWire, signal wire, or Infinite Band, which are fast enough to provide reliable communication for real-time applications. The real-time cluster is linked to one or more host stations through a Transmission Control Protocol/Internet Protocol (TCP/IP) network. Here again, the cluster of PCs can be used for real-time applications (using QNX or Red Hawk Linux), or fast simulation of complex systems (using QNX, Red Hawk, or Windows). RT-LAB PC cluster targets are designed for flexible and reconfigurable megasimulation [157]. The user can build and expand the PC

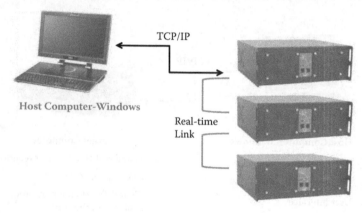

FIGURE 5.6
RT-LAB simulator with a distributed target system.

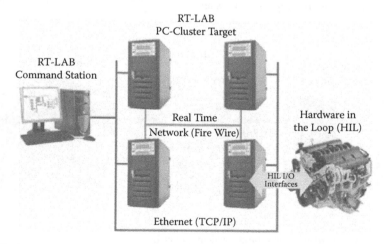

FIGURE 5.7
RT-LAB simulator with a distributed target system and HIL.

cluster as needed, and then redeploy the PCs for other applications when the simulation is done. RT-LAB can accommodate up to 64 nodes running in parallel.

5.4.3 Simulator Solvers

The RT-LAB electrical simulator uses advanced fixed-time-step solvers and computational techniques designed for the strict constraints of real-time simulation of stiff systems. They are implemented as a Simulink toolbox called ARTEMIS (Advanced Real-Time Electro-Mechanical Transient Simulator) [144], which is used with the Sim power systems. PSB (Power System Blockset) is a Simulink toolbox that enables the simulation of electric circuits and drives within the Simulink environment. While PSB now supports a fixed-time-step solver based on the Tustin method, PSB alone is not suitable for real-time simulation due to many serious limitations, including iterative calculations to solve algebraic loops, dynamic computation of circuit matrices, undamped switching oscillations, and the need for a very small step size that greatly slows down the simulation. The ARTEMIS solver uses a high-order fixed-time-step integration algorithm that is not prone to numerical oscillations, and advanced computational techniques necessary for the real-time simulation of power electronic systems and drives, such as

- Exploitation of system topology to reduce matrix size and number by splitting the equations of separated systems
- Support for parallel processing suitable for distributed simulation of large systems
- Implementation of advanced techniques for constant computation time
- Strictly noniterative integration
- Real-time compensation of switching events occurring anywhere inside the time step, enabling the use of realistic simulation step sizes while ensuring a good precision of circuits with switches (GTO, IGBT, etc.) [7, 50, 65, 98, 104, 132]

5.4.4 RT-LAB Simulation Development Procedure

Electric and power electronic systems are created on the host personal computer by interconnecting the following:

- Electrical components from component model libraries available in the Power System Blockset (PSB).
- Controller components and other components from Simulink and its toolboxes that are supported by Real-Time Workshop (RTW).

- I/O blocks from the simulator I/O toolboxes; the easy-to-use drag-and-drop Simulink interface issued at all stages of the process.
- These systems are then simulated and tuned off-line in the MATLAB/Simulink environment [144]. ARTEMIS fixed-step solvers are used for the electric part and Simulink native solvers for the controller and other block diagram parts. Finally, the model is automatically compiled and loaded to the PC cluster with the RT-LAB simulation interface [145].

The simulator software converts Simulink and Sim power systems non-real-time models to real-time simulation by providing support for

- *Model distribution*: If a model is too complex to be computed within the time step, the simulator allows the model to be distributed over several processors, automatically handling the interprocessor communication through TCP/IP, FireWire, or shared memory. Electric systems can be separated by using natural delay in the system (analog-to-digital conversion delays, filtering delays, transmission lines, etc.).
- *Multirate computation*: Not all the components in a system need to be executed at very small time steps. If the system can be separated into subsystems and executed at different update rates, cycles can be freed up for executing the subsystem(s) that need to be updated faster.
- *Specialized solvers*: The simulator uses libraries of specialized solvers (ARTEMIS) and blocks that address many of the mathematical problems that arise when taking a model to real time [140], such as new fixed-step integrators that reduce the errors introduced when replacing a variable-step integrator, and a special toolbox that compensates for errors introduced when events occur between time steps (RT events).
- *Software and hardware interfaces*: In addition to the wide range of I/O types and boards, the simulator includes a comprehensive application program interface (API) that allows signals in the model to be used in other on-line software for visualization and interaction.

5.5 PCI OP5142 Configuration

OP5142 (Figure 5.8) is one of the key building blocks in the modular OP5000 I/O system from OPAL-RT Technologies [144]. It allows the incorporation of FPGA technologies in RT-LAB simulation clusters for distributed execution of nine (hardware description language) functions and high-speed,

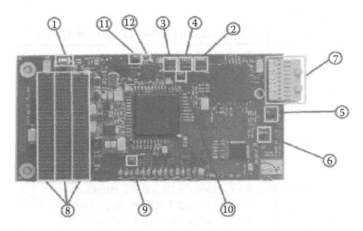

FIGURE 5.8
OP5142 layout.

high-density digital I/O in real-time models. Based on the highest-density Xilinx Spartan-3 FPGAs, the OP5142 can be attached to the backplane of an I/O module of either a Wanda 3U- or Wanda 4U-based OPAL-RT simulation system. It communicates with the target PC via a PCI Express (Peripheral Component Interconnect Express) ultra-low-latency real-time bus interface.

OP5142 includes connectivity to up to four 4U digital or analog I/O conditioning modules. This allows the incorporation of task-specific I/O hardware, such as high-speed analog signal capture and generation. Furthermore, FPGA [152] developers can incorporate their own functionality, using the System Generator for DSP toolbox or their favorite HDL development tool, through the PCI Express interface without the need for connecting to the Joint Test Action Group (JTAG) interface. Configuration files can be uploaded and stored on the built-in flash memory for instant start-up. The PCI Express port on the OP5142 adapter board allows the user to connect the distributed processors together and operate at faster cycle times than ever before. This real-time link takes advantage of the FPGA [152] power to deliver up to 2.5 Gbits/s full-duplex transfer rates. Table 5.1 gives a description of the OP5142 layout [7, 50, 65, 98, 104, 132].

5.5.1 Key Features

Reconfigurability: The OP5142 platform FPGA [152] device can be configured exactly as required by the user. Integration with Simulink, the system generator for the DSP toolbox from Xilinx, and RT-XSG from OPAL-RT Technologies allows the transfer of Simulink submodels to the OP5142 FPGA processor for distributed processing. In addition, standard and user-developed functions can be stored on the onboard flash memory for instant start-up. The OP5142 board is

TABLE 5.1

Description of Components in OP5142 Layout (OPAL-RT)

Pin	Name	Description
1	S1	FPGA Engine manual reset
2	JTAG1	FPGA JTAG interface
3	JTAG2	CPLD JTAG interface
4	JUMP4	JTAG Architecture selection
5	JTAG3	PCIe Bridge JTAG interface
6	JTAG4	SerDes JTAG interface
7	JP1	PCIe and Synchronization bus and Power supply
8	J1/J2/J3	Backplane data, ID and I2C interface
9	JUMP1	Identification EEPROM write protection
10	JUMP2	FPGA configuration mode selection
11	JUMP3	Flash memory Write protection
12	J4	Flash memory forced programmation voltage

configurable on the fly using the PCI Express bus interface and the RT-LAB design environments.

Performance: The OP5142 [7, 50, 65, 98, 104, 132] series products enable update rates of 100 MHz, providing the capability to perform time-stamped capture and generation of digital events for high-precision switching of items, such as PWM I/O signaling up to very high frequencies, as I/O scheduling is performed directly on the OP5142 board [144].

Channel density:

- Up to 256 software-configurable digital I/O lines for event capture/generation, PWM I/O, and user functions

- Up to 128 16-bit analog I/O channels, simultaneous sampling at 1 MS/s per channel for digital-to-analog conversion and 400 kS/s for analog-to-digital conversion

1. FPGA engine manual reset: This button is connected to the master reset signal of the OP5142 board. Pressing this button forces the FPGA reconfiguration, and then sends a reset signal to all OP5142 subsystems.

2. FPGA JTAG interface: This connector give access to the OP5142 JTAG chain. It is used to configure the flash memory with its default configuration file. The JTAG connection enables the user to program manually the reprogrammable components on the board and to debug the design using the chip scope, through the system generator for the DSP "chip scope" block. The use of this port is reserved for advanced users. In general, this port should not be used after the

board is manufactured. Depending upon the JUMP4 jumper presence, this interface may give access to either the FPGA and CPLD configuration or only the FPGA configuration.

3. CPLD JTAG interface: If the JUMP4 jumpers are set to the independent mode, this connector gives access to the CPLD JTAG configuration interface. The JTAG connection enables the user to program manually the reprogrammable components on the board. The use of this port is reserved for advanced users. In general, this port should not be used after the board is manufactured. If the jumpers are set to the shared mode, the CPLD and FPGA JTAG configuration are daisy-chained, and the JTAG1 connector must be used instead of this one.

4. JTAG architecture selection: This connector enables the JTAG interface of the OP5142 CPLD and FGPA to be daisy-chained. For independent operation, place the jumper between pins 6 and 8. For daisy-chain operation, place jumpers between pins 1 and 2, 3 and 4, 5 and 6, and 7 and 8, and use the FPGA JTAG connector only.

5. PCI Express bridge JTAG interface: This connector gives access to the PLX PCI Express bridge JTAG interface. It is used during manufacturing to configure the bridge with its default configuration, and should not be used by the user.

6. SerDes JTAG interface: This connector gives access to the Texas Instrument serializer–deserializer JTAG interface. It is used during manufacturing to configure the chip with its default configuration, and should not be used by the user.

7. PCI Express and synchronization bus and power supply: This port implements all data and power transfers that need to be done with the external world. It carries to the external PCI Express adapter the following:

   - The synchronization pulse train to a real-time system integration (RTSI) connector
   - Data communication packets to the PCI Express bus
   - Power supply voltages

8. Backplane data, ID, and I²C interface: These three connectors are to be attached to the Wanda Backplane Adapter J1, J2, and J3 headers. They exchange all I/O-related data to the I/O module, including identification data, serial communication with I²C devices, and user I/O data flow.

9. Identification EEPROM (Electrically Erasable Programmable Read-Only Memory) write protection: This header enables the write protection of the EEPROM located on the OP5142. EEPROM contains the board revision ID, and it always remains write protected.

10. FPGA configuration mode selection: This header enables the developer to select the way the OP5142 FPGA should be configured. The two options are (a) JTAG configuration or (b) slave parallel configuration (from the flash memory). In normal use, the FPGA should always be configured using the slave parallel feature. Note that pin 1 is on the left-hand side of the header (i.e., the "J"UMP side).

11. Flash memory write protection: This header is used to enable the developer to write some reserved sectors of the configuration flash memory. These sectors should never be used by the user.

12. Flash memory forced programmation voltage: This header provides a 12 V supply voltage to the JUMP3 connector. JUMP3 is used to enable the developer to write some reserved sectors of the configuration flash memory. These sectors should never be used by the user. Note that pin 1 is on the left-hand side of the header.

5.5.2 Technical Specifications

Digital I/O

- Number of channels: 256 input/output configurable in 1–32 bits
- Group compatibility: 3.3 V
- Power-on state: High impedance

FPGA

- Device: Xilinx Spartan 3
- I/O package: fg676
- Embedded RAM available: 216 Kbytes
- Clock: 100 MHz
- Platform options: XC3S5000
- Logic slices: 33,280
- Equivalent logic cells: 74,880
- Available I/O lines: 489

Bus

- Dimensions (not including connectors): PCI Express x1
- Data transfer: 2.5 Gbit/s

5.5.3 Analog Conversion Interface

Two types of analog conversion modules are available: the OP5340 is a bank of analog-to-digital converters (ADCs), and the OP5330 is a bank of digital-to-analog converters (DACs). The analog conversion banks must be placed onto an OP5220 passive carrier, thus providing an easy access to the modules on the front panel of the Wanda box. OP5330 and OP5340 [144] features are

- Up to 16 analog input (OP5340) or digital output (OP5330) channels
- One 16-bit ADC (OP5340) or DAC (OP5330) per channel
- Accuracy of ±5 mV
- Simultaneous sampling on all channels, which eliminates skew errors inherent in multiplexed channels
- Up to 500 kS/s update rate for every channel; total throughput of up to 8 MS/s
- Dynamic range of ±16V
- Hardware-configurable onboard signal conditioning and antialiasing filter
- Onboard EEPROM memory for calibration parameters
- Library of drag-and-drop OPAL-RT RT-XSG blocks for Simulink

5.6 System Performance of Type 2 FLC-Based p-q Control Strategy with Different Fuzzy MFs Using a Real-Time Digital Simulator

Figures 5.9 through 5.11 give the details of source voltage, load current, compensation current, source current with filter, DC link voltage, and THD of the type 2 FLC-based p-q control strategy with different fuzzy MFs under balanced, unbalanced, and nonsinusoidal supply voltage conditions using a real-time digital simulator (OPAL-RT) hardware.

Initially, the system performance is analyzed under balanced sinusoidal conditions, during which the type 2 FLC with all three MFs is good enough at suppressing the harmonics and the THDs of the p-q control strategy using a real-time digital simulator are 1.36%, 1.02%, and 0.73%, respectively. However, under unbalanced and nonsinusoidal conditions, the type 2 FLC with Gaussian MF shows superior performance. The THD of the p-q control strategy using type 2 FLC with trapezoidal MF under unbalanced condition is 2.28%, and under nonsinusoidal conditions, it is 3.31%. The THD of the p-q strategy using type 2 FLC with triangular MF under the unbalanced condition is 1.87%, and under the nonsinusoidal condition it is 2.92%. The THD of the p-q control strategy using type 2 FLC with Gaussian MF under the unbalanced condition and using a real-time digital simulator is 1.25%, and under the nonsinusoidal condition it is 2.14%.

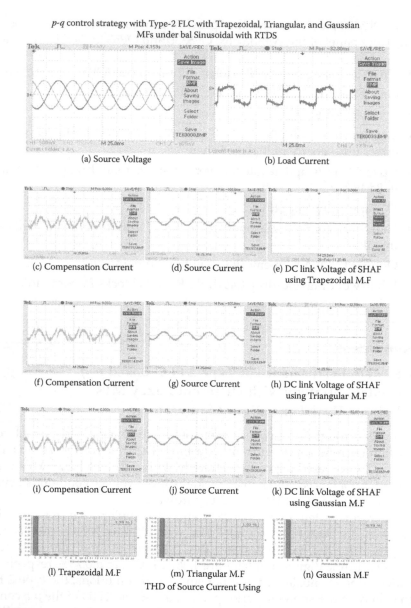

p-q control strategy with Type-2 FLC with Trapezoidal, Triangular, and Gaussian
MFs under bal Sinusoidal with RTDS

(a) Source Voltage   (b) Load Current

(c) Compensation Current   (d) Source Current   (e) DC link Voltage of SHAF
                                                          using Trapezoidal M.F

(f) Compensation Current   (g) Source Current   (h) DC link Voltage of SHAF
                                                          using Triangular M.F

(i) Compensation Current   (j) Source Current   (k) DC link Voltage of SHAF
                                                          using Gaussian M.F

(l) Trapezoidal M.F   (m) Triangular M.F   (n) Gaussian M.F
                      THD of Source Current Using

FIGURE 5.9
SHAF response using the *p-q* control strategy with type 2 FLC (trapezoidal, triangular, and
Gaussian MF) under the balanced sinusoidal condition using a real-time digital simulator.
(a) Source voltage, (b) load current (scale 30 A/div), (c) compensation current (scale 20 A/div)
using trapezoidal MF, (d) source current (scale 40 A/div) with filter using trapezoidal MF, (e) DC
link voltage using trapezoidal MF, (f) compensation current using triangular MF, (g) source
current with filter using triangular MF, (h) DC link voltage using triangular MF, (i) compensation current using Gaussian MF, (j) source current with filter using Gaussian MF, (k) DC link
voltage using Gaussian MF, (l) THD of source current with trapezoidal MF, (m) THD of source
current with triangular MF, and (n) THD of source current with Gaussian MF.

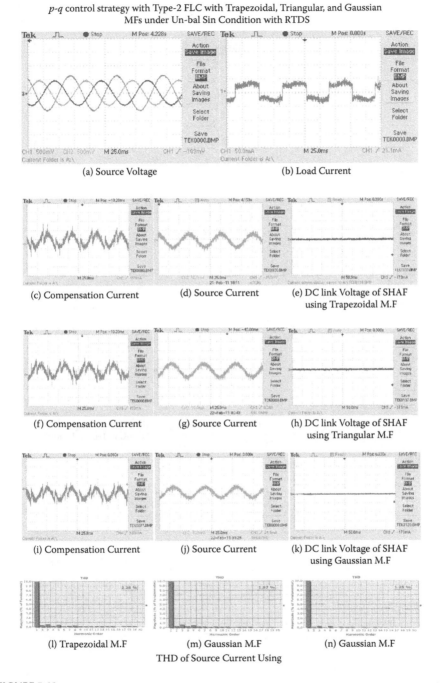

p-q control strategy with Type-2 FLC with Trapezoidal, Triangular, and Gaussian
MFs under Un-bal Sin Condition with RTDS

(a) Source Voltage   (b) Load Current

(c) Compensation Current   (d) Source Current   (e) DC link Voltage of SHAF
using Trapezoidal M.F

(f) Compensation Current   (g) Source Current   (h) DC link Voltage of SHAF
using Triangular M.F

(i) Compensation Current   (j) Source Current   (k) DC link Voltage of SHAF
using Gaussian M.F

(l) Trapezoidal M.F   (m) Gaussian M.F   (n) Gaussian M.F
THD of Source Current Using

FIGURE 5.10
SHAF response using the *p-q* control strategy with type 2 FLC (trapezoidal, triangular, and
Gaussian MF) under the unbalanced sinusoidal condition using a real-time digital simulator.

p-q control strategy with Type-2 FLC with Trapezoidal, Triangular, and Gaussian
MFs under Nonsinusoidal with RTDS

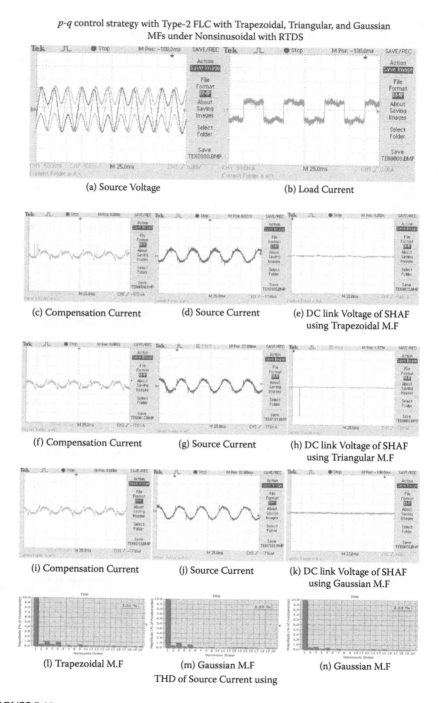

(a) Source Voltage

(b) Load Current

(c) Compensation Current

(d) Source Current

(e) DC link Voltage of SHAF
using Trapezoidal M.F

(f) Compensation Current

(g) Source Current

(h) DC link Voltage of SHAF
using Triangular M.F

(i) Compensation Current

(j) Source Current

(k) DC link Voltage of SHAF
using Gaussian M.F

(l) Trapezoidal M.F

(m) Gaussian M.F

(n) Gaussian M.F

THD of Source Current using

FIGURE 5.11
SHAF response using the *p-q* control strategy with type 2 FLC (trapezoidal, triangular, and
Gaussian MF) under the nonsinusoidal condition using a real-time digital simulator.

Even though the *p-q* control strategy using type 2 FLC with different fuzzy MFs is able to mitigate the harmonics, a small amount of notches are observed in the source current. The main reason behind the notches is that the controller failed to track the current correctly, and thereby APF fails to compensate completely. So to mitigate the harmonics perfectly, one has to choose the perfect control strategy. So to avoid the difficulties that occur with the *p-q* control strategy, we have considered the I_d-I_q control strategy with type 2 FLC.

5.7 System Performance of Type 2 FLC-Based I_d-I_q Control Strategy with Different Fuzzy MFs Using a Real-Time Digital Simulators

Figures 5.12 through 5.14 give the details of source voltage, load current, compensation current, source current with filter, DC link voltage, and THD of the type 2 FLC-based I_d-I_q control strategy with different fuzzy MFs using a real-time digital simulator.

Initially, the system performance is analyzed under balanced sinusoidal conditions, during which the type 2 FLC with all three MFs (triangular, trapezoidal, and Gaussian) is good enough at suppressing the harmonics. The respective THDs of SHAF in the real-time digital simulator are 0.66%, 0.52%, and 0.43%.

The THD of the I_d-I_q control strategy using type 2 FLC with trapezoidal MF under the unbalanced condition is 1.38%, and under the nonsinusoidal condition it is 2.62%. The THD of the I_d-I_q control strategy using type 2 FLC with triangular MF under the unbalanced condition is 1.17%, and under the nonsinusoidal condition it is 1.98%. The THD of the I_d-I_q control strategy using type 2 FLC with Gaussian MF under the unbalanced condition is 0.79%, and under the nonsinusoidal condition it is 1.63%. These are all obtained using a real-time digital simulator.

While considering the I_d-I_q control strategy using type 2 FLC with different fuzzy MFs, SHAF succeeds in compensating harmonic currents. It is observed that source current waveforms are very good; notches in the waveform are eliminated by using the I_d-I_q control strategy with type 2 FLC different fuzzy MFs.

The real-time implementation results showed that even if the supply voltage is nonsinusoidal, the performance with the I_d-I_q theory with type 2 FLC comfortably outperformed the results obtained using the *p-q* theory with type 2 FLC.

I_d-I_q control strategy with Type-2 FLC with Trapezoidal, Triangular, and Gaussian MFs under Balanced Sinusoidal

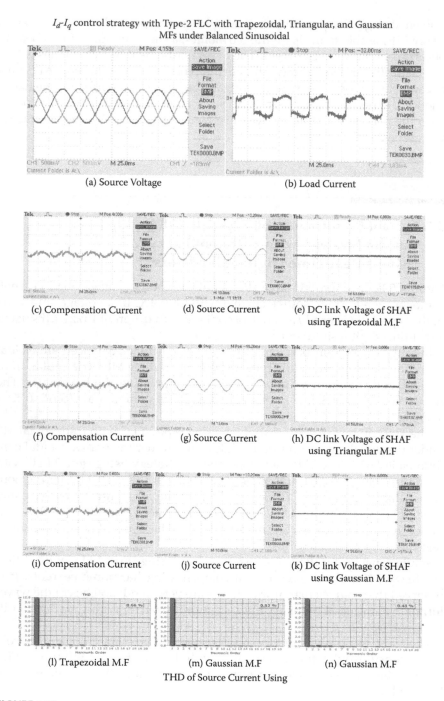

(a) Source Voltage

(b) Load Current

(c) Compensation Current

(d) Source Current

(e) DC link Voltage of SHAF using Trapezoidal M.F

(f) Compensation Current

(g) Source Current

(h) DC link Voltage of SHAF using Triangular M.F

(i) Compensation Current

(j) Source Current

(k) DC link Voltage of SHAF using Gaussian M.F

(l) Trapezoidal M.F

(m) Gaussian M.F

(n) Gaussian M.F

THD of Source Current Using

FIGURE 5.12

SHAF response using the I_d-I_q control strategy with type 2 FLC (trapezoidal, triangular, and Gaussian MF) under the balanced sinusoidal condition using a real-time digital simulator.

I_d-I_q control strategy with Type-2 FLC with Trapezoidal, Triangular, and Gaussian
MFs under Un-bal Sinusoidal

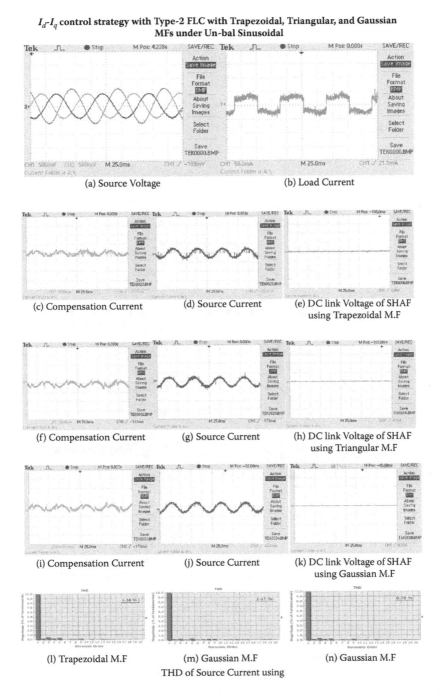

(a) Source Voltage

(b) Load Current

(c) Compensation Current

(d) Source Current

(e) DC link Voltage of SHAF
using Trapezoidal M.F

(f) Compensation Current

(g) Source Current

(h) DC link Voltage of SHAF
using Triangular M.F

(i) Compensation Current

(j) Source Current

(k) DC link Voltage of SHAF
using Gaussian M.F

(l) Trapezoidal M.F

(m) Gaussian M.F

(n) Gaussian M.F

THD of Source Current using

FIGURE 5.13
SHAF response using the I_d-I_q control strategy with type 2 FLC (trapezoidal, triangular, and Gaussian MF) under the unbalanced sinusoidal condition using a real-time digital simulator.

I_d-I_q control strategy with Type-2 FLC with Trapezoidal, Triangular, and Gaussian
MFs under Nonsinusoidal with RTDS

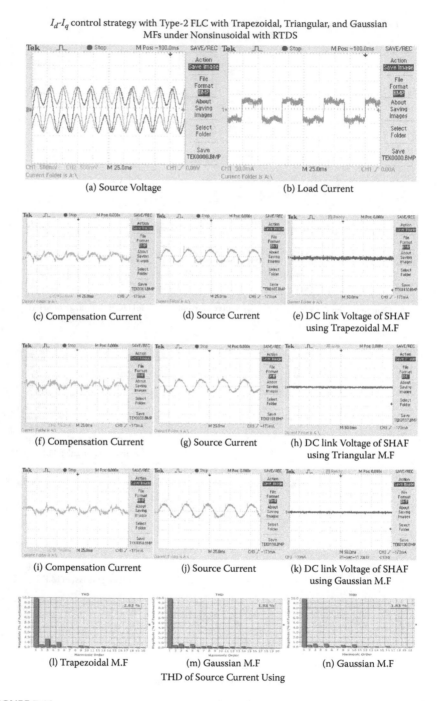

(a) Source Voltage

(b) Load Current

(c) Compensation Current

(d) Source Current

(e) DC link Voltage of SHAF
using Trapezoidal M.F

(f) Compensation Current

(g) Source Current

(h) DC link Voltage of SHAF
using Triangular M.F

(i) Compensation Current

(j) Source Current

(k) DC link Voltage of SHAF
using Gaussian M.F

(l) Trapezoidal M.F

(m) Gaussian M.F

(n) Gaussian M.F

THD of Source Current Using

FIGURE 5.14
SHAF response using the I_d-I_q control strategy with type 2 FLC under the nonsinusoidal condition using a real-time digital simulator.

5.8 Comparative Study

Figures 5.15 and 5.16 clearly illustrate the THD of the source current for the p-q and I_d-I_q control strategies using the PI controller, type 1 FLC, and type 2 FLC with different fuzzy MFs using a real-time digital simulator under various source conditions.

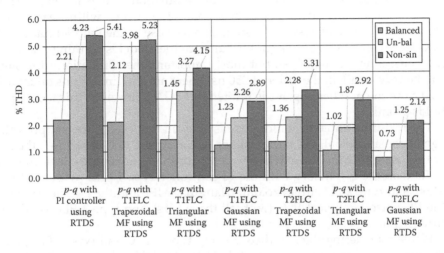

FIGURE 5.15

Bar graph indicating the THD of the source current for the p-q control strategy using the PI controller, type 1 FLC, and type 2 FLC with different fuzzy MFs using a real-time digital simulator.

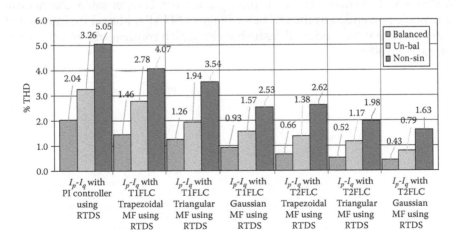

FIGURE 5.16

Bar graph indicating the THD of the source current for the I_d-I_q control strategy using the PI controller, type 1 FLC, and type 2 FLC with different fuzzy MFs using a real-time digital simulator.

5.9 Summary

Modern power systems continue to evolve, requiring constant evaluation of new constraints. Major studies will require the use of very fast, flexible, and scalable real-time simulators. This chapter has introduced a specific class of digital simulator known as a real-time simulator. The latest trend in real-time simulation consists of exporting simulation models to FPGA. This approach has many advantages. First, computation time within each time step is almost independent of system size because of the parallel nature of FPGAs. Second, overruns cannot occur once the model is running and timing constrains are met. Last, but most important, the simulation time step can be very small, in the order of 250 ns. There are still limitations on model size since the number of gates is limited in FPGAs. However, this technique holds promise.

The real-time implementation results showed that even if the supply voltage is nonsinusoidal, the performance of the type 2 FLC-based I_d-I_q theory with Gaussian MF showed better compensation capabilities in terms of THD than the I_d-I_q theory with the PI controller, type 1 FLC (with all three MFs), and type 2 FLC (trapezoidal, triangular MF), and the p-q theory with the PI controller, type 1 FLC (with all three MFs), and type 2 FLC (with all three MFs).

While considering the I_d-I_q theory using type 1 FLC and type 2 FLC, and the p-q control strategy with type 2 FLC, the SHAF has been found to meet the IEEE 519–1992 recommendations on harmonic levels, making it easily adaptable to more severe constraints, such as highly distorted and unbalanced supply voltage. The control approach has compensated the neutral harmonic currents, and the DC bus voltage of SHAF is almost maintained at the reference value under all disturbances, which confirms the effectiveness of the controller.

6

Conclusions and Future Scope

6.1 Conclusions

In this book, with the main objective being compensation of current harmonics generated due to the presence of nonlinear loads in a three-phase four-wire distribution system, the research studies presented start with an introduction to harmonics, clearly specifying its description, causes, and consequences. The role of passive power filters in harmonics elimination is discussed. But to avoid the inevitable drawbacks of passive filters, we moved toward the use of active power filters (APFs). After comparing various APF configurations, the three-phase four-wire capacitor midpoint shunt APF VSI-PWM (voltage source inverter–pulse width modulation) configuration was chosen for the modeling of shunt APF.

The performances of the p-q and I_d-I_q control strategies were evaluated by comparing the total harmonic distortions (THDs) in compensated source currents and DC link voltage regulation under balanced, unbalanced, and distorted/nonsinusoidal supply conditions. DC link voltage regulation with the help of the proportional–integral (PI) controller, type 1 fuzzy logic controller (FLC), and type 2 FLC with different fuzzy MFs (trapezoidal, triangular, and Gaussian MF) to minimize the power losses occurring inside APF is studied. The various drawbacks encountered in the conventional PI controller were discussed. Next, research concentrated on the implementation of type 1 FLC with different fuzzy MFs (trapezoidal, triangular, and Gaussian MF); this also suffers from several drawbacks, resulting in severe deterioration of APF performance. Hence, we developed type 2 FLC with different fuzzy MFs that could overcome the drawbacks observed in the PI and type 1 FLC. Three-phase reference current waveforms generated by the proposed scheme were tracked by the three-phase voltage source converter in a hysteresis band control scheme. The performance of the control strategies was evaluated in terms of harmonic mitigation and DC link voltage regulation. The simulation (MATLAB®) results were validated with real-time implementation on a real-time digital simulator (OPAL-RT).

The investigations carried out in this book yield the following important conclusions:

- It can be inferred from the simulation and real-time results of Chapter 2 that the p-q control strategy yields inadequate results under unbalanced and nonsinusoidal source voltage conditions.

- Under unbalanced and nonsinusoidal conditions, the p-q control strategy does not succeed in compensating harmonic currents; notches are observed in the source current. The main reason behind the notches is that the controller failed to track the current correctly, and thereby APF fails to compensate completely. So to avoid the difficulties occurring with the p-q control strategy, we considered the I_d-I_q control strategy.

- The I_d-I_q scheme is the best APF control scheme for compensation of current harmonics for a wide variety of supply voltage and loading conditions. The THD in source current can also be satisfactorily lowered below 5%, thereby satisfying the IEEE 519 standards on a harmonic level. Simultaneously, it also compensates for excessive neutral current.

- Under unbalanced and nonsinusoidal conditions, the PI controller is unable to maintain a constant DC link voltage (V_{dc} is nearer to 780 V, but V_{dc-ref} is 800 V), and it is unable to mitigate the harmonics completely and THD is close to 5%. The mitigation of harmonics is poor when the THD of the source current is greater. But according to IEEE 519–1992, THD must be less than 5%. So to mitigate harmonics effectively, we considered type 1 FLC with different fuzzy MFs.

- With the I_d-I_q control strategy using type 1 FLC with different fuzzy MFs, the SHAF is able to maintain a constant DC link voltage (V_{dc} is nearer to 790 V, but V_{dc-ref} is 800 V), and it is able to mitigate harmonics (THD is nearly equal to 2.5%–3.5%) in a better way than the p-q control strategy using type 1 FLC with different fuzzy MFs (THD is nearly equal to 3%–5%).

- Even though the I_d-I_q control strategy using type 1 FLC with different fuzzy MFs is able to mitigate the harmonics, notches are present in the source current. So to avoid the difficulties that occur with type 1 FLC-based p-q and I_d-I_q control strategies with different fuzzy MFs, we considered type 2 FLC with different fuzzy MFs.

- The proposed type 2 FLC-based SHAF with different fuzzy MFs is able to eliminate the uncertainty in the system, and SHAF gains outstanding compensation abilities. Type 2 FLC is able to maintain a constant DC link voltage (V_{dc} is nearer to 797 V, which is almost equal to V_{dc-ref} 800 V), and it is able to mitigate harmonics (THD is nearly equal to 1%–2%) in a superior way compared to type 1 FLC (THD is nearly equal to 2.5%–5%).

- While considering the I_d-I_q control strategy using type 1 FLC (trapezoidal, triangular, and Gaussian MF) and type 2 FLC (trapezoidal, triangular, and Gaussian MF) and the p-q control strategy with type 1 FLC (triangular and Gaussian MF) and type 2 FLC (trapezoidal, triangular, and Gaussian MF), the SHAF has been found to meet the IEEE 519–1992 standard recommendations on harmonic levels.

- The PI controller–, type 1 FLC–, and type 2 FLC–based shunt active filter control strategies (p-q and I_d-I_q) with different fuzzy MFs (trapezoidal, triangular, and Gaussian) are verified with real-time digital simulator (OPAL-RT) hardware to validate the proposed research.

- The simulation and real-time implementation results showed that even if the supply voltage is unbalanced or nonsinusoidal, the performance of SHAF using the I_d-I_q theory with type 2 FLC (Gaussian MF) shows better compensation capabilities in terms of THD than the I_d-I_q theory with PI, type 1 FLC (trapezoidal, triangular, and Gaussian MF), and type 2 FLC (trapezoidal, triangular MF), and the p-q theory with PI, type 1 FLC (trapezoidal, triangular, and Gaussian MF), and type 2 FLC (trapezoidal, triangular and Gaussian MF). The control approach compensates the neutral harmonic currents, and the DC link voltage of SHAF is almost maintained at the reference value under all disturbances, which confirms the effectiveness of the controller.

- RT-LAB real-time simulation results further confirm the results obtained from MATLAB simulations; that is, the I_d-I_q scheme is the best reference compensation current extraction scheme, and the type 2 FLC-based controller is the most efficient out of all the other conventional (PI) and type 1 FLC-based controllers discussed in this book.

The findings of the above investigations are summarized in Figures 6.1 through 6.4. These figures clearly illustrate the amount of THD of source current reduced from one controller to another for shunt active filter control strategies under various source conditions using MATLAB and a real-time digital simulator.

6.2 Future Scope

In this book, efforts have been made to improve the power quality of power systems by mitigating the current harmonics and maintaining a constant DC link voltage using PI controller–, type 1 FLC–, and type 2 FLC–based shunt active filter control strategies (p-q and I_d-I_q) with different fuzzy MFs under

148

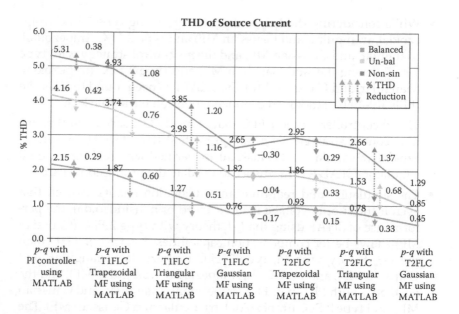

FIGURE 6.1

Line graph indicating the amount of THD reduced for the PI controller, type 1 FLC, and type 2 FLC with different fuzzy MFs using the *p-q* control strategy with MATLAB.

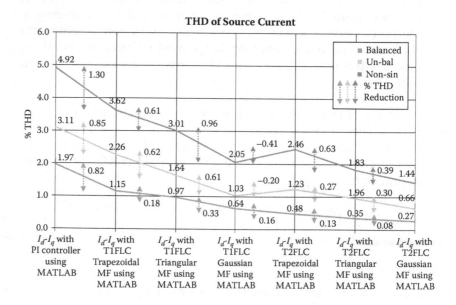

FIGURE 6.2

Line graph indicating the amount of THD reduced for the PI controller, type 1 FLC, and type 2 FLC with different fuzzy MFs using the I_d-I_q control strategy with MATLAB.

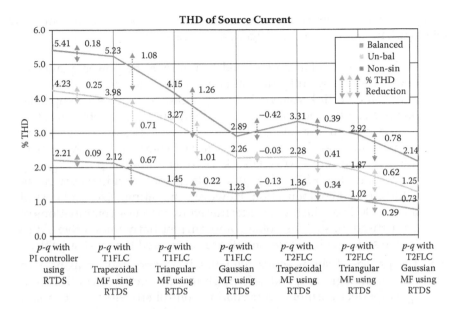

FIGURE 6.3
Line graph indicating the amount of THD reduced for the PI controller, type 1 FLC, and type 2 FLC with different fuzzy MFs using the *p-q* control strategy with a real-time digital simulator.

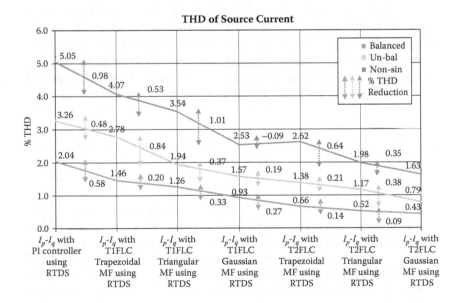

FIGURE 6.4
Line graph indicating the amount of THD reduced for the PI controller, type 1 FLC, and type 2 FLC with different fuzzy MFs using the I_d-I_q control strategy with a real-time digital simulator.

various source voltage conditions. It has explored some good ideas and suitable solutions, but further investigation is necessary.

- In this book, we have attempted to improve the power quality of power systems by mitigating the current harmonics and maintaining a constant DC link voltage. Many applications require a compensation of a combination of voltage- and current-based problems, a few of them being interrelated. A hybrid of active series with active shunt filters is an ideal choice for such a mixed compensation. Moreover, this hybrid of both APFs (also known as a unified power quality conditioner (UPQC)) is also quite suitable for individual current- or voltage-based compensation. However, the rating, size, and cost of this UPQC is on the higher side; therefore, for a few combinations of compensation, such as voltage and current harmonics, other APFs (active series with passive shunt) are considered the most suitable.

- The proposed shunt active filter has been operated by using two control strategies: p-q and I_d-I_q. Applying the improved control strategies (perfect harmonic cancellation control strategy, etc.) can still improve the power quality. So, the potential of the proposed version could be explored by adopting different control strategies.

- The proposed shunt active filter has been operated by using three controllers—PI controller, type 1 FLC, and type 2 FLC—with different fuzzy MFs. Applying the improved controllers (neurofuzzy, etc.) can still improve the power quality. So, the potential of the proposed version could be explored by adopting different controllers.

- In this book, the hysteresis current control scheme was considered for generation of gating signals to the devices of the APF. Applying the improved switching techniques (adaptive hysteresis controller, etc.) can still improve the output quality. So, the potential of the proposed version could be explored by adopting different switching techniques.

- In this book, we considered the Mamdani type of fuzzy logic controller with 49 rules. The types of membership functions are trapezoidal, triangular, and Gaussian. There are seven membership functions. The type of implication is the Mamdani max-min operation, and the type of defuzzification method is the centroid of area method. Takai Sugeno fuzzy logic control (TS FLC) is better than the Mamdani type of fuzzy control in the sense that it requires only 2 fuzzy sets and 4 rules compared to the 7 fuzzy sets and 49 rules used for the Mamdani type FLC. Hence, the TS FLC is a good aspirant for improving the performance of a system by eliminating the harmonics.

References

1. H. Akagi, New trends in active filters for power conditioning, *IEEE Trans. Ind. Appl.*, 32(6), 1312–1322, 1996.
2. H. Rudnick, J. Dixon, and L. Morán, Active power filters as a solution to power quality problems in distribution networks, *IEEE Power & Energy Magazine*, September/October 2003), 32–40.
3. S. Bhattacharya and D. M. Divan, Hybrid series active/parallel passive power line conditioner with controlled harmonic injection, U.S. Patent 5465203, November 1995.
4. B. Singh, K. Al-Haddad, and A. Chandra, A review of active filters for power quality improvement, *IEEE Trans. Ind. Electron.*, 46(5), 1999.
5. J. H. Choi, G. W. Park, and S. B. Dewan, Standby power supply with active power filter ability using digital controller, in *Proceedings of IEEE APEC '95*, 1995), 783–789.
6. Z. Li, H. Jin, and G. Joos, Control of active filters using digital signal processors, in *Proceedings of IEEE IECON '95*, 1995), 651–655.
7. S. Mikkili and A. K. Panda, Real-time implementation of PI and fuzzy logic controllers based shunt active filter control strategies for power quality improvement, *Int. J. Electr. Power Energy Syst.*, 43, 1114–1126, 2012.
8. A. Teke, L. Saribulut, and M. Tumay, A novel reference signal generation method for power quality improvement of unified power quality conditioner, *IEEE Trans. Power Delivery*, 26(4), 2205–2214, 2011.
9. S. Saad and L. Zellouma, Fuzzy logic controller for three-level shunt active filter compensating harmonics and reactive power, *Electric Power Syst. Res.*, 79, 1337–1341, 2009.
10. P. Kumar and A. Mahajan, Soft computing techniques for the control of an active power filter, *IEEE Trans. Power. Delivery*, 24(1), 2009.
11. F. Mekri, B. Mazari, and M. Machmoum, Control and optimization of shunt active power filter parameters by fuzzy logic, *Can. J. Elect. Comput. Eng.*, 31(3), 2006.
12. P. Kirawanich and R. M. O'Connell, Fuzzy logic control of an active power line conditioner, *IEEE Trans. Power. Electron.*, 19(6), 2004.
13. S. K. Jain, P. Agrawal, and H. O. Gupta, Fuzzy logic controlled shunt active power filter for power quality improvement, *IEEE Proc. Electric Power Appl.*, 149(5), 2002.
14. A. Bhattacharya and C. Chakraborty, A shunt active power filter with enhanced performance using ANN-based predictive and adaptive controllers, *IEEE Trans. Ind. Electron.*, 58(2), 2011.
15. M. Cirrincione, M. Pucci, and G. Vitale, A single-phase DG generation unit with shunt active power filter capability by adaptive neural filtering, *IEEE Trans. Ind. Electron.*, 55(5), 2008.
16. H. Cheng Lin, Intelligent neural network-based fast power system harmonic detection, *IEEE Trans. Ind. Electron.*, 54(1), 2007.

17. D. Ould Abdeslam, P. Wira, J. Merckle, D. Flieller, and Y.-A. Chapuis, A unified artificial neural network architecture for active power filters, *IEEE Trans. Ind. Electron.*, 54(1), 2007.
18. F. Javier Alcantara and Patricio Salmeron, A new technique for unbalance current and voltage estimation with neural networks, *IEEE Trans. Power Syst.*, 20(2), 2005.
19. J. R. Vazquez and P. Salmeron, Active power filter control using neural network technologies, *IEE Proc. Electric Power Appl.*, 150(2), 2003.
20. S. M. R. Rafiei, R. Ghazi, and H. A. Toliyat, IEEE-519-based real-time and optimal control of active filters under non-sinusoidal line voltages using neural networks, *IEEE Trans. Power Delivery*, 17(3), 2002.
21. Y.-M. Chen and R. M. O'Connell, Active power line conditioner with a neural network control, *IEEE Trans. Ind. Appl.*, 33(4), 1997.
22. M. Rukonuzzaman and M. Nakaoka, Single-phase shunt active power filter with harmonic detection, *IEE Proc. Electric Power Appl.*, 149(5), 2002.
23. A. C. Liew, Excessive neutral currents in three-phase fluorescent lighting circuits, *IEEE Trans. Ind. Appl.*, 25, 776–782, 1989.
24. T. M. Gruzs, A survey of neutral currents in three-phase computer power systems, *IEEE Trans. Ind. Appl.*, 26, 719–725, 1990.
25. F. Z. Peng, G. W. Ott Jr., and D. J. Adams, Harmonic and reactive power compensation based on the generalized instantaneous reactive power theory for three-phase four-wire systems, *IEEE Trans. Power Electron.*, 13(5), 1174–1181, 1998.
26. S. Mikkili and A. K. Panda, PI and fuzzy logic controller based 3-phase 4-wire shunt active filter for mitigation of current harmonics with I_d-I_q control strategy, *J. Power Electron.*, 11(6), 2011.
27. A. Mansoor, W. M. Grady, P. T. Staats, R. S. Thallam, M. T. Doyle, and M. J. Samotyj, Predicting the net harmonic currents produced by large numbers of distributed single-phase computer loads, *IEEE Trans. Power Delivery*, 10, 2001–2006, 1994.
28. Active filters, Technical document, 2100/1100 series, Mitsubishi Electric Corp., Tokyo, 1989, pp. 1–36.
29. A. H. Kikuchi, Active power filters, in Toshiba GTR module (IGBT) application notes, Toshiba Corp., Tokyo, 1992, pp. 44–45.
30. S. A. Moran and M. B. Brennen, Active power line conditioner with fundamental negative sequence compensation, U.S. Patent 5384696, January 1995.
31. G. T. Heydt, *Electric Power Quality*, Stars in a Circle, West Lafayette, IN, 1991.
32. D. D. Shipp, Harmonic analysis, suppression for electrical systems supplying static power converters other nonlinear loads, *IEEE Trans. Ind. Appl.*, 15, 453–458, 1979.
33. L. Rossetto and P. Tenti, Evaluation of instantaneous power terms in multiphase systems: Techniques, application to power-conditioning equipment, *Eur. Trans. Elect. Power Eng.*, 4(6), 469–475, 1994.
34. L. S. Czarnecki, Combined time-domain, frequency-domain approach to hybrid compensation in unbalanced nonsinusoidal systems, *Eur. Trans. Elect. Power Eng.*, 4(6), 477–484, 1994.
35. A. Chaoui, F. Krim, J.-P. Gaubert, and L. Rambault, DPC controlled three-phase active filter for power quality improvement, *Int. J. Electr. Power Energy Syst.*, 30, 476–485, 2008.

36. N. Zaveri and A. Chudasama, Control strategies for harmonic mitigation and power factor correction using shunt active filter under various source voltage conditions, *Int. J. Electr. Power Energy Syst.*, 42, 661–671, 2012.

37. H. Akagi, Y. Kanazawa, and A. Nabae, Instantaneous reactive power compensators comprising switching devices without energy storage components. *IEEE Trans. Ind. Appl.*, Ia-20(3), 625–630, 1984.

38. H. Akagi et al., *Instantaneous Power Theory and Applications to Power Conditioning*, IEEE Press/Wiley-Interscience, 2007.

39. H. Akagi, Y. Kanazawa, and A. Nabae, Generalized theory of the instantaneous reactive power in three-phase circuits, in *Proceedings of IPEC Tokyo*, 1983, pp. 1375–1386.

40. A. Ferrero and G. S. Furga, A new approach to the definition of power components in three-phase systems under non-sinusoidal conditions, *IEEE Trans. Instrum. Meas.*, 40, 568–577, 1991.

41. H. Akagi and A. Nabae, The p-q theory in three-phase systems under non-sinusoidal conditions, *Eur. Trans. Elect. Power Eng.*, 3(1), 27–31, 1993.

42. Z. Zhou and Y. Liu, Pre-sampled data based prediction control for active power filters, *Int. J. Electr. Power Energy Syst.*, 37, 13–22, 2012.

43. V. Soares, P. Verdelho, and G. D. Marques, An instantaneous active and reactive current component method for active filters, *IEEE Trans. Power Electron.*, 15(4), 660–669, 2000.

44. V. Soares et al., Active power filter control circuit based on the instantaneous active and reactive current i_d-i_q method, in *IEEE Power Electronics Specialists Conference*, 1997, pp. 1096–1101.

45. J. L. Willems, Current compensation in three-phase power systems, *Eur. Trans. Elect. Power Eng.*, 3(1), 61–66, 1993.

46. B. Singh, P. Jayaprakash, and D. P. Kothari, New control approach for capacitor supported DSTATCOM in three-phase four wire distribution system under non-ideal supply voltage conditions based on synchronous reference frame theory, *Int. J. Electr. Power Energy Syst.*, 33, 1109–1117, 2011.

47. P. Rodriguez, J. I. Candela, A. Luna, and L. Asiminoaei, Current harmonics cancellation in three-phase four-wire systems by using a four-branch star filtering topology, *IEEE Trans. Power Electron.*, 24(8), 1939–1950, 2009.

48. M. I. M. Montero, E. R. Cadaval, and F. B. Gonzalez, Comparison of control strategies for shunt active power filters in three-phase four wire systems, *IEEE Trans. Ind. Electron.*, 22(1), 229–236, 2007.

49. P. Salmeron and R. S. Herrera, Distorted and unbalanced systems compensation within instantaneous reactive power framework, *IEEE Trans. Power Delivery*, 21(3), 1655–1662, 2006.

50. S. Mikkili and A. K. Panda, Simulation and real-time implementation of shunt active filter i_d-i_q control strategy for mitigation of harmonics with different fuzzy membership functions, *IET Power Electron.*, 5(9), 1856–1872, 2012.

51. I. Holh, Pulse width modulation: A survey, *IEEE Trans. Ind. Electron.*, 39(5), 410–420, 1999.

52. A. Albanna and C. J. Hatziadoniu, Harmonic analysis of hysteresis controlled grid-connected inverters, in *Proceedings of IEEE/PES Power Systems Conference and Exposition (PSCE '09)*, 2009, pp. 1–8.

53. A. Karaarslan, Hysterisis control of power factor correction with a new approach of sampling technique, in *Proceedings of IEEE 25th Convention of Electrical and Electronics Engineers in Israel*, 2008, pp. 765–769.

54. F. Mekri, M. Machmoum, N. Ait Ahmed, and B. Mazari, A fuzzy hysteresis voltage and current control of an unified power quality conditioner, in *Proceedings of 34th Annual Conference of IEEE IECON*, 2008, pp. 2684–2689.

55. H. Sasaki and T. Machida, A new method to eliminate AC harmonic currents by magnetic flux compensation: Considerations on basic design, *IEEE Trans. Power App. Syst.*, PAS-90, 2009–2019, 1971.

56. Harmonic currents, static VAR systems, Inform. NR500–015E, ABB Power Systems, Stockholm, Sweden, September 1988, pp. 1–13.

57. H. Akagi, S. Atoh, and A. Nabae, Compensation characteristics of active power filter using multi-series voltage-source PWM converters, *Elect. Eng. Japan*, 106(5), 28–36, 1986.

58. S. Fukuda and M. Yamaji, Design, characteristics of active power filter using current source converter, in *Conference Record of IEEE-IAS Annual Meeting*, 1990, pp. 965–970.

59. C. K. Duffey and R. P. Stratford, Update of harmonic standard IEEE-519: IEEE recommended practices, requirements for harmonic control in electric power systems, *IEEE Trans. Ind. Appl.*, 25, 1025–1034, 1989.

60. B. N. Singh, B. Singh, A. Chandra, and K. Al-Haddad, DSP based implementation of sliding mode control on an active filter for voltage regulation and compensation of harmonics, power factor and unbalance of nonlinear loads, in *IEEE IECON '99*, 1999, vol. 2, pp. 855–860.

61. F. Ferreira, L. Monteiro, J. L. Afonso, and C. Couto, A control strategy for a three-phase four-wire shunt active filter, in *IEEE IECON*, 2008, pp. 411–416.

62. J. Dixon, J. Contardo, and L. Moran, DC link fuzzy control for an active power filter, sensing the line current only, in *Proceedings of 28th Annual IEEE Power Electronics Specialists Conference (PESC '97)*, 1997, vol. 2, pp. 1109–1114.

63. M Suresh, A. K. Panda, S. S. Patnaik, and S. Yellasiri, Comparison of two compensation control strategies for shunt active power filter in three-phase four-wire system, in *Proceedings of IEEE PES Innovative Smart Grid Technologies (ISGT)*, Hilton Anaheim, CA, 2011, pp. 1–6. DOI: 10.1109/ISGT.2011.5759126.

64. A. D. Le Roux, J. A. Du Toit, and J. H. R. Enslin, Integrated active rectifier and power quality compensator with reduced current measurement, *IEEE Trans. Ind. Electron.*, 46(3), 504–511, 1999.

65. S. Mikkili and A. K. Panda, RTDS hardware implementation and simulation of SHAF for mitigation of harmonics using p-q control strategy with PI and fuzzy logic controllers, *Front. Electr. Electron. Eng.*, 7(4), 427–437, 2012.

66. W. V. Lyon, Reactive power and unbalanced circuits, *Electr. World*, 75(25), 1417–1420, 1920.

67. C. I. Budeanu, *Puissances reactives et fictives*, pub. 2, Institute Romaine de Energy, Bucharest, 1927.

68. C. I. Budeanu, The different options and conceptions regarding active power in non-sinusoidal systems, pub. 4, Institute Romaine de Energy, Bucharest, 1927.

69. S. Fryze, Wirk, Blind und Scheinleistung in Elektrischen Stromkreisen mit nicht-sinusformigen verlauf von storm and Spannung, *ETZ-Arch. Electrotech.*, 53, 596–599, 625–627, 700–702, 1932.

70. M. S. Erlicki and A. E. Eigeles, New aspects of power factor improvements: Part I: Theoretical basis, *IEEE Trans. Ind. Gen. Appl.*, 1GA-4, 441–446, 1968.

71. H. Sasaki and T. Machida, New method to eliminate AC harmonic by magnetic compensation: Consideration on basic design, *IEEE Trans. Power App. Syst.*, 90(5), 2009–2019, 1970.

72. T. Fukao, H. Ilda, and S. Miyairi, Improvements of the power factor of distorted waveforms by thyristor based switching filter, *Trans. IEE Japan B*, 92(6), 342–349, 1970.

73. L. Gyugyi and B. R. Pelly, *Static Power Frequency Chargers: Theory Performance and Application*, Wiley, New York, 1976.

74. F. Harashima, H. Inaba, and K. Tsuboi, A closed-loop control system for the reduction of reactive power required by electronic converters, *IEEE Trans. IECI*, 23(2), 162–166, 1976.

75. L. Gyugyi and E. C. Strycula, Active power filters, *in Proceedings of IEEE Industrial Application Annual Meeting*, 1976, vol. 19-C, pp. 529–535.

76. I. Takahashi, K. Fuziwara, and A. Nabae, Universal reactive power compensator, in *Proceedings of IEEE Industrial Application Annual Meeting*, 1980, pp. 858–863.

77. I. Takahashi, K. Fuziwara, and A. Nabae, Distorted current compensation system using thyristor based line commutated converters, *Trans. IEE Japan B*, 101(3), 121–128, 1981.

78. H. Akagi, Y. Kanazawa, and A. Nabae, Principles and compensation effectiveness of instantaneous reactive power compensation devices, in *Meeting of the Power Semiconductor Converters Researchers—IEE Japan*, 1982, SPC-82-16.

79. H. Akagi, Y. Kanazawa, and A. Nabae, Generalized theory of the instantaneous reactive power and its applications, *Trans. IEE Japan B*, 103(7), 483–490, 1983.

80. E. Clarke, Circuit analysis of AC power systems, in *Symmetrical and Related Components*, Wiley, New York, 1943.

81. F. M. Uriarte, Hysterisis modeling by inspection, in *Proceedings of 38th North American Power Symposium*, 2006, pp. 187–191.

82. L. A. Zadeh, Fuzzy sets, *Inf. Control*, 8, 338–353, 1965.

83. L. A. Zadeh, Outline of a new approach to the analysis of complex systems and decision processes, *IEEE Trans. Syst. Man Cybern.*, 3(1), 28–44, 1973.

84. E. H. Mamdani, Applications of fuzzy logic to approximate reasoning using linguistic synthesis, *IEEE Trans. Comput.*, 26(12), 1182–1191, 1977.

85. E. H. Mamdani, Advances in the linguistic synthesis of fuzzy controllers, *Int. J. Man Mach. Stud.*, 8, 669–678, 1976.

86. J. Zhao and B. K. Bose, Evaluation of membership functions for fuzzy logic controlled induction motor drive, in *28th Annual Conference of IECON*, 2002, vol. 1, pp. 229–234.

87. B. K. Bose, *Modern Power Electronics and AC Drives*, Prentice Hall, Upper Saddle River, NJ, 2002.

88. L. A. Zadeh, Fuzzy logic, *Computer*, 21(4), 83–93, 1988.

89. A. Kandel, *Fuzzy Expert Systems*, CRC Press, Boca Raton, FL, 1992.

90. C. C. Lee, Fuzzy logic in control systems: Fuzzy logic controller: Part 1, *IEEE Trans. Syst. Man Cybern.*, 20(2), 404–418, 1990.

91. C. C. Lee, Fuzzy logic in control systems: Fuzzy logic controller: Part 2, *IEEE Trans. Syst. Man Cybern.*, 20(2), 419–435, 1990.

92. B. Kosko, *Neural Networks and Fuzzy Systems: A Dynamical Systems Approach*, Prentice Hall, Upper Saddle River, NJ, 1991.

93. Fuzzy inference system, http://www.mathworks.in/help/toolbox/fuzzy/fp351dup8.html.
94. T. A. Runkler, Selection of appropriate defuzzification methods using application specific properties, *IEEE Trans. Fuzzy Syst.*, 5(1), 72–79, 1997.
95. Defuzzification methods, http://www.mathworks.in/products/demos/shipping/fuzzy/defuzzdm.html.
96. Fuzzy inference system editor, http://www.mathworks.in/products/fuzzy-logic/description3.html.
97. D. Dubois and H. Prade, *Fuzzy Sets and Systems: Theory and Applications*, Academic Press, New York, 1980.
98. S. Mikkili and A. K. Panda, Real-time implementation of power theory using FLC based shunt active filter with different fuzzy M.F.s, in *IECON 2012: 38th Annual Conference on IEEE Industrial Electronics Society*, Montreal, QC, October 25–28, 2012, pp. 702–707.
99. L. A. Zadeh, Toward a restructuring of the foundations of fuzzy logic (FL), in *IEEE International Conference on World Congress on Computational Intelligence Fuzzy Systems*, 1998, vol. 2, pp. 1676–1677.
100. L. A. Zadeh, Fuzzy logic: Issues, contentions and perspectives, in *IEEE International Conference on Acoustics, Speech, and Signal Processing*, 1994, vol. 6.
101. W. Pedrycz and F. Gomide, The design of fuzzy sets, in *Fuzzy Systems Engineering: Toward Human-Centric Computing*, IEEE Press, NJ, 2007, pp. 67–100.
102. A. Elmitwally, S. Abdelkader, and M. Elkateb, Performance evaluation of fuzzy controlled three and four wire shunt active power conditioners, in *IEEE Power Engineering Society Winter Meeting*, 2000, vol. 3, pp. 1650–1655.
103. J.-S. R. Jang, C.-T. Sun, and E. Mizutani, *Neuro-Fuzzy and Soft Computing*, Prentice-Hall, Englewood Cliffs, NJ, 1997.
104. S. Mikkili and A. K. Panda, Type-1 and type-2 FLC based shunt active filter I_d-I_q control strategy with different fuzzy MFs for power quality improvement using RTDS hardware, *IET Power Electron.* Vol. 6(4), 818–833, 2013.
105. W. Pedrycz and F. Gomide, Operations and aggregations of fuzzy sets, in *Fuzzy Systems Engineering: Toward Human-Centric Computing*, IEEE Press, 2007, pp. 101–138.
106. M. Kantardzic, Fuzzy sets and fuzzy logic, in *Data Mining: Concepts, Models, Methods, and Algorithms*, Wiley-IEEE Press, NJ, 2011, pp. 414–446.
107. S. Yeong Yi and M. Jin Chung, Robustness of fuzzy logic control for an uncertain dynamic system, *IEEE Trans. Fuzzy Syst.*, 6(2), 1998.
108. A. Celikyilmaz and I. B. Turksen, Uncertainty modelling of improved fuzzy functions with evolutionary systems, *IEEE Trans. Syst. Man Cybern. B*, 38(4), 1098–1110, 2008.
109. J. M. Mendel, Uncertainty, fuzzy logic, and signal processing, *Signal Proc. J.*, 80, 913–933, 2000.
110. L. A. Zadeh, From fuzzy logic to extended fuzzy logic: A first step, in *IEEE Annual Meeting of the North American Fuzzy Information Processing Society*, 2009, pp. 1–2.
111. R. Sepulveda, O. Castillo, P. Melin, A. Rodriguez-Diaz, and O. Montiel, Handling uncertainties in controllers using type-2 fuzzy logic, in *Proceedings of IEEE FUZZ Conference*, Reno, NV, May 2005, pp. 248–253.

112. R. Sepulveda, O. Castillo, P. Melin, A. Rodriguez-Diaz, and O. Montiel, Experimental study of intelligent controllers under uncertainty using type-1 and type-2 fuzzy logic, *Inf. Sci.*, 177, 2023–2048, 2007.
113. O. Castillo and P. Melin, *Recent Advances in Interval Type-2 Fuzzy Systems*, Springer Briefs in Applied Sciences and Technology, vol. 1, Springer, Berlin, 2012. DOI: 10.1007/978-3-642-28956-9.
114. M. Mizumoto and K. Tanaka, Some properties of fuzzy sets of type-2, *Inf. Control*, 31, 312–340, 1976.
115. L. A. Zadeh, The concept of a linguistic variable and its application to approximate reasoning: Parts 1, 2, and 3, *Inf. Sci.*, 8, 199–249; 8, 301–357; 9, 43–80, 1975.
116. O. Castillo and P. Melin, *Type-2 Fuzzy Logic Theory and Applications*, Springer Verlag, Berlin, 2008.
117. N. N. Karnik and J. M. Mendel, An introduction to type-2 fuzzy logic systems, University of Southern California, Los Angeles, June 1998b.
118. N. N. Karnik, J. M. Mendel, and Q. Liang, Type-2 fuzzy logic systems, *IEEE Trans. Fuzzy Syst.*, 7, 643–658, 1999.
119. J. M. Mendel, *Uncertain Rule-Based Fuzzy Logic Systems: Introduction and New Directions*, Prentice-Hall, Upper Saddle River, NJ, 2001.
120. J. M. Mendel, R. I. John, and F. Liu, Interval type-2 fuzzy logic systems made simple, *IEEE Trans. Fuzzy Syst.*, 14, 808–821, 2006.
121. W. Dongrui and J. M. Mendel, On the continuity of type-1 and interval type-2 fuzzy logic systems, *IEEE Trans. Fuzzy Syst.*, 19(1), 179–192, 2011.
122. Q. Liang and J. Mendel, Interval type-2 fuzzy logic systems: Theory and design, *IEEE Trans. Fuzzy Syst.*, 8, 535–550, 2000.
123. J. Castro, O. Castillo, and L. G. Martínez, Interval type-2 fuzzy logic toolbox, *Eng. Lett.*, 15(1), EL_15_1_14, 2007.
124. J. Castro, O. Castillo, and P. Mellin, A type-2 fuzzy logic toolbox for control applications, in *Proceedings of IEEE FUZZ Conference*, London, July 2007, pp. 61–66.
125. E. H. Mamdani and S. Assilian, An experiment in linguistic synthesis with a fuzzy logic controller, *Int. J. Man Mach. Stud.*, 7(1), 1–13, 1975.
126. S. Coupland and R. I. John, Geometric type-1 and type-2 fuzzy logic systems, *IEEE Trans. Fuzzy Syst.*, 15, 3–15, 2007.
127. G. J. Klir and B. Yuan, *Fuzzy Sets and Fuzzy Logic: Theory and Applications*, Prentice Hall, Upper Saddle River, NJ, 1995.
128. L. A. Zadeh, Knowledge representation in fuzzy logic, *IEEE Trans. Knowl. Data Eng.*, 1, 89–100, 1989.
129. L. A. Zadeh, Fuzzy logic computing with words, *IEEE Trans. Fuzzy Syst.*, 2, 103–111, 1996.
130. R. I. John and S. Coupland, Type-2 fuzzy logic: A historical view, *IEEE Comput. Intell. Mag.*, 2, 57–62, 2007.
131. H. Hagras, Type-2 FLCs: A new generation of fuzzy controllers, *IEEE Comput. Intell. Mag.*, 2, 30–43, 2007.
132. S. Mikkili and A. K. Panda, Performance analysis and real-time implementation of shunt active filter current control strategy with type-1 and type-2 FLC triangular M.F, *Int. Trans. Elect. Eng. Sys.*, 24(3), 347–362, March 2014.
133. R. R. Yager, On a general class of fuzzy connectives, *J. Fuzzy Sets Syst.*, 4, 235–242, 1980.

134. R. R. Yager, A characterization of the fuzzy extension principle, *J. Fuzzy Sets Syst.*, 18, 205–217, 1986.

135. J. R. Aguero and A. Vargas, Calculating function of interval type-2 fuzzy numbers for fault current analysis, *IEEE Trans. Fuzzy Syst.*, 15, 31–40, 2007.

136. D. Dubois and H. Prade, Operations on fuzzy numbers, *Int. J. Syst. Sci.*, 9, 613–626, 1978.

137. N. N. Karnik and J. M. Mendel, Operations on type-2 fuzzy sets, *Fuzzy Sets Syst.*, 122, 327–348, 2001.

138. J. M. Mendel, Type-2 fuzzy sets and systems: An overview, *IEEE Comput. Intell. Mag.*, 2, 20–29, 2007.

139. S. Coupland and R. John, A fast geometric method for defuzzification of type-2 fuzzy sets, *IEEE Trans. Fuzzy Syst.*, 16(4), 929–941, 2008.

140. D. Dubois and H. Prade, Operations in a fuzzy-valued logic, *Inf. Control*, 43, 224–240, 1979.

141. R. R. Yager, On the implication operator in fuzzy logic, *Inf. Sci.*, 31, 141–164, 1983.

142. A. Kaufman and M. M. Gupta, *Introduction to Fuzzy Arithmetic: Theory and Applications*, Van Nostrand Reinhold, New York 1991.

143. L. X. Wang, *Adaptive Fuzzy Systems and Control: Design and Stability Analysis*, Prentice Hall, Upper Saddle River, NJ, 1994.

144. RT-LAB Professional, http://www.opal-rt.com/product/rt-lab-professional.

145. Real time simulation, en.wikipedia.org/wiki/Real-time_Simulation.

146. M. Papini and P. Baracos, Real-time simulation, control and HIL with COTS computing clusters, presented at AIAA Modeling and Simulation Technologies Conference, Denver, CO, August 2001.

147. C. Dufour, G. Dumur, J. N. Paquin, and J. Belanger, A multicore PC-based simulator for the hardware-in-the-loop testing of modern train and ship traction systems, in *13th Power Electronics and Motion Control Conference*, Poland, 2008, pp. 1475–1480.

148. D. Auger, Programmable hardware systems using model-based design, in *IET and Electronics Weekly Conference on Programmable Hardware Systems*, London, 2008, pp. 1–12.

149. V. Q. Do, J.-C. Soumagne, G. Sybille, G. Turmel, P. Giroux, G. Cloutier, and S. Poulin, Hypersim, an integrated real-time simulator for power networks and control systems, in *ICDS '99*, Vasteras, Sweden, May 1999, pp. 1–6.

150. J. Belanger, P. Venne, and J. N. Paquin, The what, where and why of real-time simulation. http://www.opal-rt.com/sites/default/files/technical_papers/ PES-GM-Tutorial_04%20-%20Real%20Time%20Simulation.pdf, pp. 37–49.

151. S. Abourida, C. Dufour, J. Bélanger, and V. Lapointe, Real-time, PC-based simulator of electric systems and drives, presented at *International Conference on Power Systems Transients—IPST 2003*, New Orleans, 2003.

152. M. Matar and R. Iravani, FPGA implementation of the power electronic converter model for real-time simulation of electromagnetic transients, *IEEE Trans. Power Delivery*, 25(2), 852–860, 2010.

153. J. A. Hollman and J. R. Marti, Real time network simulation with PC-cluster, *IEEE Trans. Power Syst.*, 18(2), 563–569, 2008.

154. I. Etxeberria-Otadui, V. Manzo, S. Bacha, and F. Baltes, Generalized average modelling of FACTS for real time simulation in ARENE, in *IEEE 28th Annual Conference of the Industrial Electronics Society (IECON '02)*, 2002, vol. 2, pp. 864–869.

155. J.-F. Cecile, L. Schoen, V. Lapointe, A. Abreu, and J. Belanger, A Distributed Real-Time Framework for Dynamic Management of Heterogeneous Co-stimulations, 2006.
156. G. Sybille and H. Le-Huy, Digital simulation of power systems and power electronics using the MATLAB/Simulink Power System Block set, in *IEEE Power Engineering Society Winter Meeting 2000*, Singapore, 2000, 4), 2973–2981.
157. J. Belanger, V. Lapointe, C. Dufour, and L. Schoen, eMEGAsim: An open high-performance distributed real-time power grid simulator: Architecture and specification, presented at International Conference on Power Systems (ICPS '07), Bangalore, India, December 12–14, 2007.

Index

Milton Keynes UK
Ingram Content Group UK Ltd.
UKHW040054071024
449327UK00019B/566